CONTENTS

INTRODUCTION

The word peristalsis is derived from the Greek word $\pi\varepsilon\rho\iota\sigma\tau\alpha\lambda\tau\iota kos$, which means clasping and compressing. It is used to describe a progressive wave of contraction along a channel or tube whose cross-sectional area consequently varies. In physiology, peristalsis is used by the body to propel or mix the contents of a tube as in uretra, gastro-intestinal tract, bile ducts and other glandular ducts. Some worms to make locomotion using the mechanism of peristalsis. Roller and finger pumps using viscous fluids also operate on this principle. Peristalsis has been proposed as a mechanism for the transport of spermatozoa in vas deferens Semans (1938), which connects the ducts. The principle of peristaltic transport has been exploited for industrial applications like sanitary fluid transport, blood pumps in heart lungs machine and transport of corrosive fluids where the contact of the fluid with the machinery parts is prohibited. Since the first investigation of Latham (1966), a number of analytic, numerical and experimental studies of peristaltic flow of different fluids have been reported under different conditions with reference to physiological and mechanical situations. A numerical technique using boundary integral method has been developed by Pozrikidis (1987) to investigate peristaltic transport in an asymmetric channel under Stokes flow conditions to understand the fluid dynamics involved. He has studied the streamline patterns and mean flow rate due to different amplitudes and phases of the wall deformation. The existence of trapping regions adjacent to the walls is also observed for some flow rates.

Recently, physiologists observed that the intra uterine fluid flow due to myometrial contractions is peristaltic-type motion and the myometrial contractions may occur in both symmetric and asymmetric directions, De varies et al. (1990). Eytan et al. (1999) have observed that the characterization of non-pregnant woman contractions is very complicated as they are composed of variable amplitudes, a range of frequencies and different wave lengths. It was observed that the width of the sagittal cross-section of the uterine cavity increases towards the fundus and the cavity is not fully occluded during the contractions. In view of this, Eytan and Elad (1999) have developed a mathematical model of wall induced peristaltic intra-uterine fluid flow in a two-dimensional channel with wave trains having a phase

3

difference moving independently on the upper and lower walls. These results have been used to evaluate fluid flow pattern in a non-pregnant uterus. They have also calculated the possible particle trajectories to understand the transport of embryo before it gets implanted at the uterine-wall. Ramachandra Rao and Mishra (2004) discussed the peristaltic transport of a Newtonian fluid in an asymmetric channel. Abd Elnaby and Haroun (2006) presented a new model to study the effect of wall properties as peristaltic transport of a viscous fluid. Subba Reddy *et al.* (2007) investigated peristaltic motion of a power-law fluid in an asymmetric channel. Vajravelu *et al.* (2009) examined peristaltic transport of a Casson fluid in contact with a Newtonian fluid in a circular tube with permeable wall. The effects of wall permeability and yield stress on the pumping characteristics have been reported in their investigation.

Due to the flow behaviour of non-Newtonian fluids the governing equations become more complex to handle as additional non-linear terms appear in the equations of motion. There is also no universal constitutive model available which exhibits the characteristics of the all non-Newtonian fluids. Mention may be made to some interesting studies done previously, pertaining to non-Newtonian fluids, which may give insights into their behaviour. Some recent studies have been made on the peristaltic motions of conducting, Newtonian and non-Newtonian fluids in asymmetric channels. The MHD flow of a fluid in a channel with elastic, rhythmically contracting walls is of interest in connection with certain problems of the movement of conductive physiological fluids, e.g., the blood and with the need for theoretical research on the operation of a peristaltic MHD compressor. The effect of a moving magnetic field on blood flow was studied by Stud *et al.* (1977), and they observed that the effects of a suitable moving magnetic field accelerate the speed of blood. Srivastava and Agarwal (1980) considered the blood as an electrically conducting fluid that constitutes a suspension of red cells in the plasma. Mekheimer (2008) analyzed the MHD flow of a conducting couple stress fluid in a slit channel with rhythmically contracting walls. Srinivas and Kothandapani (2008) have analyzed the MHD peristaltic flow of a viscous fluid in an asymmetric channel with heat transfer. Kothadapani and Srinivas (2008) have examined the influence of wall properties in the MHD peristaltic transport with heat transfer and porous medium. Wang *et al.* (2008) have studied the MHD peristaltic motion of a Sisko fluid in an asymmetric channel and Kothandapani and Srinivas (2008) have examine the peristaltic transport of a Jeffrey fluid under the effect of magnetic field in asymmetric channel with flexible rigid walls. In view of these facts, it will be interesting to study the peristaltic flow of conducting Jeffrey fluid flow in a channel bounded by permeable walls.

In the present analysis the fluid considered is of Jeffrey type and is electrically conducting. The Jeffrey model is a relatively simpler linear model using time derivatives instead of convective derivatives and it represents a rheology different from the Newtonian. The main purpose of the present study is to investigate the peristaltic pumping of MHD flow of a Jeffrey fluid in a two-dimensional asymmetric channel having electrically insulated walls. The channel asymmetry is produced by choosing the peristaltic wave train on the walls which have different amplitudes and phase due to the variation in channel width, wave amplitudes and phase differences. The governing equations of fluid flow are solved subject to relevant boundary conditions. The comparison among the three wave forms is also made carefully and influence of several pertinent parameters on the stream function and pressure drop have been studied and numerical results obtained are presented. The results and discussions presented in this study may be helpful to further understanding MHD peristaltic motion for non-Newtonian fluids in asymmetric channels.

1.1   Mathematical formulation

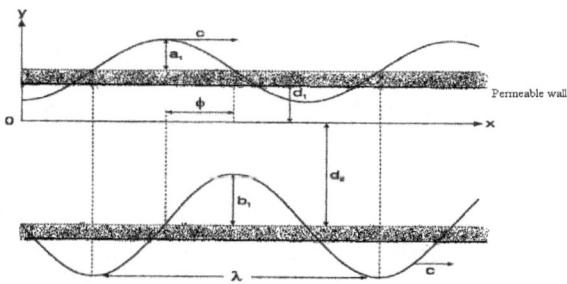

Figure1.1. Schematic diagram of a two-dimensional asymmetric channel

We consider the motion of an incompressible viscous fluid in a two – dimensional channel included by individual wave trains propagating with constant speed c along the permeable walls of the channel. The wall deformations are given by

$$\overline{h_1}(\overline{X},\overline{t}) = d_1 + a_1 \cos\left(\frac{2\pi}{\lambda}(\overline{X} - c\overline{t})\right)\dots\text{upper wall,} \tag{1.1}$$

$$\overline{h_2}(\overline{X},\overline{t}) = d_2 + b_1 \cos\left(\frac{2\pi}{\lambda}(\overline{X} - c\overline{t}) + \phi\right)\dots\text{lower wall,} \tag{1.2}$$

where a_1, b_1 are the amplitude of the waves, λ is the wavelength, $d_1 + d_2$ (Figure 1.1) is the width of the channel, the phase difference ϕ varies in the range of $0 \le \phi \le \pi, \phi = 0$, corresponds to symmetric channel with waves out of phase and $\phi = \pi$ the waves are in phase and further a_1, b_1, d_1, d_2 and ϕ satisfies the condition

$$a_1^2 + b_1^2 + 2a_1 b_1 \cos\phi \le (d_1^2 + d_2^2)^2. \tag{1.3}$$

1.3 Equations of motion

The constitutive equations for an incompressible Jeffrey fluid are

$$\overline{T} = \overline{p}\,\overline{I} + \overline{S}, \tag{1.4}$$

$$\overline{S} = \frac{\mu}{1+\lambda_1}(\dot{\overline{r}} + \lambda_2 \ddot{\overline{r}}). \tag{1.5}$$

where $\overline{T}$ and $\overline{S}$ are Cauchy stress tensor and extra stress tensor, respectively, $\overline{p}$ is the pressure, $\overline{I}$ is the identity tensor, λ_1 is the ratio of relaxation to retardation times, λ_2 is the retardation time $\ddot{r}$ is the shear rate and dots over the quantities indicate differentiation with respect to time.

In laboratory frame, the equation governing two–dimensional motion of an incompressible, MHD Jeffrey fluid are

$$\frac{\partial \overline{U}}{\partial X} + \frac{\partial \overline{V}}{\partial Y} = 0, \tag{1.6}$$

$$\rho\left[\frac{\partial}{\partial t} + \overline{U}\frac{\partial}{\partial x} + \overline{V}\frac{\partial}{\partial y}\right]\overline{U} = -\frac{\partial \overline{p}}{\partial x} + \frac{\partial(\overline{S_{xx}})}{\partial x} + \frac{\partial(\overline{S_{xy}})}{\partial Y} - \sigma B_0^2 \overline{U}, \tag{1.7}$$

6

$$\rho\left[\frac{\partial}{\partial t}+\overline{U}\frac{\partial}{\partial x}+\overline{V}\frac{\partial}{\partial y}\right]\overline{V}=-\frac{\partial\overline{p}}{\partial x}+\frac{\partial(\overline{S}_{xx})}{\partial x}+\frac{\partial(\overline{S}_{yy})}{\partial Y}. \tag{1.8}$$

where $\overline{U},\overline{V}$ are the velocity components in the laboratory frame $(\overline{X},\overline{Y})$, ρ is the density, μ is the coefficient of viscosity of the fluid, $\overline{p}$ is the pressure and σ is the electrical conductivity of the fluid. We shall carry out their investigation in a coordinate system moving with the wave speed, in which the boundary shape is stationary.

The coordinates and velocities in the laboratory frame $(\overline{X},\overline{Y})$ and the wave frame $(\overline{x},\overline{y})$ are related by

$$\overline{X}=\overline{x}-ct, \overline{y}=\overline{Y}, \overline{u}=\overline{U}-c, \overline{v}=\overline{V}, \overline{p}(x)=\overline{P}(X,t).$$

where $\overline{u}, \overline{v}$ are the velocity components in the wave frame $\overline{p}$ and $\overline{P}$ are pressures in wave and fixed frame of references, respectively.

Introducing the following non–dimensional quantities:

$$x=\frac{2\pi\overline{x}}{\lambda}, \; y=\frac{\overline{y}}{d_1}, \; u=\frac{\overline{u}}{c}, \; v=\frac{\overline{v}}{c\delta}, \; \delta=\frac{2\pi d_1}{\lambda}, \; p=\frac{2\pi d_1^2\overline{P}}{\mu c\lambda}, \; t=\frac{2\pi c\overline{t}}{\lambda},$$

$$h_1=\frac{\overline{h_1}}{d_1}, \; h_2=\frac{\overline{h_2}}{d_1}, \; S=\frac{d_1\overline{S}}{\mu c}, \; d=\frac{d_2}{d_1}, \; a=\frac{a_1}{d_1}, \; b=\frac{b_1}{d_1}, \; \mathrm{Re}=\frac{\rho c d_1}{\mu}.$$

and the stream function

$$u=\frac{\partial\psi}{\partial y}, \; v=-\delta\frac{\partial\psi}{\partial x}. \tag{1.9}$$

in Navier–Stokes equation and eliminating pressure by cross differentiation, we get

$$\delta\mathrm{Re}\left[\left(\frac{\partial\psi}{\partial y}\frac{\partial}{\partial x}-\frac{\partial\psi}{\partial x}\frac{\partial}{\partial y}\right)\Delta^2\psi\right]=\left[\left(\frac{\partial^2}{\partial y^2}-\delta^2\frac{\partial^2}{\partial x^2}\right)S_{xy}\right]+\partial\left[\frac{\partial^2}{\partial x\partial y}(S_{xx}-S_{xy})\right]-M^2\frac{\partial^2\psi}{\partial y^2}. \tag{1.10}$$

in which

$$S_{xx}=\frac{2\partial}{1+\lambda_1}\left[1+\frac{\delta\lambda_2 c}{d_1}\left(\frac{\partial\psi}{\partial y}\frac{\partial}{\partial y}-\frac{\partial\psi}{\partial x\partial y}\right)\right]\frac{\partial^2\psi}{\partial x\partial y}, \tag{1.11}$$

$$S_{xy}=\frac{1}{1+\lambda_1}\left[1+\frac{\delta\lambda_2 c}{d_1}\left(\frac{\partial\psi}{\partial y}\frac{\partial}{\partial y}-\frac{\partial\psi}{\partial x\partial y}\right)\right]\left(\frac{\partial^2\psi}{\partial y^2}-\delta^2\frac{\partial^2\psi}{\partial x^2}\right), \tag{1.12}$$

7

$$S_{yy} = \frac{2\delta}{1+\lambda_1}\left[1+\frac{\delta\lambda_2 c}{d_1}\left(\frac{\partial\psi}{\partial y}\frac{\partial}{\partial y}-\frac{\partial\psi}{\partial x\partial y}\right)\right]\frac{\partial^2\psi}{\partial x\partial y},$$

(1.13)

$$\Delta^2 = \left(\delta^2\frac{\partial^2\psi}{\partial x^2}+\frac{\partial^2\psi}{\partial y^2}\right),$$

where $M = \sqrt{\frac{\sigma}{\mu}}B_0 d_1$ is the Hartmann number.

Using the long wave length approximation and neglecting the wave number along with low-Reynolds number, are can find from equations (1.10) to (1.13) that

$$\frac{\partial^2}{\partial y^2}\left[\frac{1}{1+\lambda^2}\frac{\partial^2\psi}{\partial y^2}\right]-M^2\frac{\partial^2\psi}{\partial y^2}=0.$$

(1.14)

The dimensionless boundary conditions are

$$\psi = \frac{q}{2} \text{ at } y = h_1 = 1+a\cos 2\pi x,$$

(1.15)

$$\psi = -\frac{q}{2} \text{ at } y = h_2 = -d-b\cos(2\pi x+\phi),$$

(1.16)

$$\frac{\partial\psi}{\partial y}+\beta\frac{\partial^2\psi}{\partial y^2}=-1 \text{ at } y = h_1,$$

(1.17)

$$\frac{\partial\psi}{\partial y}-\beta\frac{\partial^2\psi}{\partial y^2}=-1 \text{ at } y = h_2.$$

(1.18)

where q is the flux, $\beta = \frac{\sqrt{k}}{\alpha d_1} = \frac{\sqrt{Da}}{\alpha}$ is the permeability parameter including slip,

Da is the Darcy number, α is the slip parameter and k is the permeability.

The boundary conditions (1.17) and (1.18) are introduced at the permeable walls of the channel following Saffman (1971). When the flow takes place past a permeable bed, it is well known that the usual no-slip condition is not valid at the nominal surface of the porous bed (Beavers and Joseph, 1967). The slip at the permeable wall is presented through a slip condition formulated by Saffman which is an improved condition to the Beavers and Joseph slip condition.

in wave frame and a, b, ϕ and d satisfy the relation

8

$$a^2 + b^2 + 2ab\cos\phi \le \left(1+d\right)^2.$$

1.4 Solution of the problem

In order to apply the Adomian decomposition method, equation (1.14) can be written as

$$L\psi = N\left(\psi_m\right)_{yy}, \tag{1.19}$$

where $N^2 = M^2(1+\lambda_1)$ and $L = \dfrac{d^4}{dy^4}$. Since a fourth-order difference operator, L^{-1} is a fourth-fold integration operator defined by

$$L^{-1} = \int_0^y\int_0^\tau\int_0^\eta\int_0^\xi (.)\,d\xi\,d\eta\,d\tau\,dy. \tag{1.20}$$

Operating with L^{-1}, equation (1.19) yields

$$\psi = c_1 + c_2 y + c_3 \frac{y^2}{2!} + c_4 \frac{y3}{3!} + NL^{-1}(\psi_m)_{yy} \tag{1.21}$$

in which the function c_i (i=1 to 4) can be determined by utilizing the boundary conditions (1.15) to (1.18).

By the standard Adomian decomposition method, one can write $\psi = \displaystyle\sum_{m=0}^{\infty} \psi_m$. (1.22)

where the components $\psi_m, m \ge 0$, will be determine recursively. From equations (1.20) and (1.21), we obtain the following recursively relation

$$\psi_0 = c_1 + c_2 y + c_3 \frac{y^2}{2!} + c_4 \frac{y3}{3!}, \tag{1.23}$$

$$\psi_{m+1} = L^{-1} N(\psi_m)_{yy}, \ m \ge 0. \tag{1.24}$$

hence

$$\psi_1 = \frac{1}{N^3} c_3 \frac{(Ny)^4}{4!} + \frac{1}{N^4} c_4 \frac{(Ny)^5}{5!},$$

$$\psi_2 = \frac{1}{N^3} c_3 \frac{(Ny)^6}{6!} + \frac{1}{N^4} c_4 \frac{(Ny)^7}{7!},$$

$$\psi_m = \frac{1}{N^3}c_3\frac{(Ny)^{2m+2}}{(2m+2)!}+\frac{1}{N^4}c_4\frac{(Ny)^{2m+3}}{(2m+3)!}, \quad m \geq 0.$$

Through equation (1.22), the expression for ψ is easily seen to have the form

$$\psi = c_1 + c_2 y + \frac{1}{N^3}c_3(\cosh Ny - 1) + \frac{1}{N^4}c_4(\sinh Ny - Ny), \qquad (1.25)$$

which may be simplified as

$$\psi = F_1 + F_2 y + F_3 \cosh Ny + F_4 \sinh Ny. \qquad (1.26)$$

The velocity is given

$$u = F_2 + NF_3 \sinh Ny + NF_4 \cosh Ny. \qquad (1.27)$$

where the values of F_1 to F_4 can be found by using the boundary conditions (1.15) to (1.18) and are given by

$$F_1 = \frac{q}{2}+\frac{Nqh_1 \cosh N(\frac{h_1-h_2}{2})+\left[(2+N^2q\beta)h_1-(q+h_1-h_2)\right]\sinh N(\frac{h_1-h_2}{2})}{\left[2-N^2\beta(h_1-h_2)\right]\sinh N(\frac{h_1-h_2}{2})-N(h_1-h_2)\cosh N(\frac{h_1-h_2}{2})},$$

$$F_2 = -\frac{\left[(2+N^2q\beta)\sinh N(\frac{h_1-h_2}{2})+Nq\cosh N(\frac{h_1-h_2}{2})\right]}{\left[(2-N^2\beta(h_1-h_2))\sinh N(\frac{h_1-h_2}{2})\right]-N(h_1-h_2)\cosh N(\frac{h_1-h_2}{2})},$$

$$F_3 = -\frac{\left[(q+h_1-h_2)\sinh N(\frac{h_1+h_2}{2})\right]}{[2-N^2\beta(h_1-h_2)]\sinh N(\frac{h_1-h_2}{2})-N(h_1-h_2)\cosh N(\frac{h_1-h_2}{2})},$$

$$F_4 = \frac{(q+h_1-h_2)\cosh N(\frac{h_1+h_2}{2})}{\left[2-N^2\beta(h_1-h_2)\right]\sinh N(\frac{h_1-h_2}{2})-N(h_1-h_2)\cosh N(\frac{h_1-h_2}{2})}.$$

The flux at any axial station in the fixed frame is

$$Q = \int_{h_2}^{h_1}(\frac{\partial\psi}{\partial y}+1)dy = h_1 - h_2 + q, \qquad (1.28)$$

The average volume flow rate over one period of the peristaltic wave is defined as

$$\bar{Q} = \frac{1}{T}\int_0^T Q \, dT = \frac{1}{T}\int_0^T (q + h_1 - h_2) \, dt = q + 1 + d.$$ (1.29)

The pressure gradient is obtained from the dimensionless momentum equation for the axial velocity as

$$\frac{\partial p}{\partial x} = \frac{N^3}{1+\lambda_1}\left[\frac{(q+h_1-h_2)\cosh N\frac{(h_1-h_2)}{2}}{\left[2-N^2\beta(h_1-h_2)\right]\sinh N\frac{(h_1-h_2)}{2} - N(h_1-h_2)\cosh N\frac{(h_1-h_2)}{2}}\right].$$ (1.30)

The non–dimensional expression for the pressure rise ΔP_λ is given as follows

$$\Delta P_\lambda = \int_0^{2\pi} (\frac{\partial p}{\partial x}) \, dx.$$ (1.31)

The non-dimensional shear stress at the upper wall of the channel reduces to

$$S_{xy} = \frac{1}{1+\lambda_1}\frac{\partial^2 \psi}{\partial y^2},$$

$$= \frac{1}{1+\lambda_1}\left[\frac{N^2(q+h_1-h_2)\left[(\sinh N\left(\frac{h_1+h_2}{2}\right)\cosh Ny + \cosh n\left(\frac{h_1+h_2}{2}\right)\sinh Ny\right]}{\left[2-N^2(h_1-h_2)\beta\right]\sinh N\left(\frac{h_1-h_2}{2}\right)-N(h_1-h_2)\cosh N\left(\frac{h_1-h_2}{2}\right)}\right].$$ (1.32)

1.5 Expressions for wave shape:

The non-dimensional expressions for three considered wave forms are given by the following equations:

1. Sinusoidal wave

$$h(x) = 1 + a\sin(x),$$ (1.33)

2. Triangular wave

$$h(x) = 1 + a\left\{\frac{8}{\pi^3}\sum_{m=1}^{\infty}\frac{(-1)^{m+1}}{(2m-1)^2}\sin[(2m-1)x]\right\},$$ (1.34)

3. Trapezoidal wave

$$h(x)=1+a\left\{\frac{32}{\pi^2}\sum_{m=1}^{\infty}\frac{\sin\frac{\pi}{8}(2m-1)}{(2m-1)^2}\sin[(2m-1)x]\right\}. \qquad (1.35)$$

1.6 Results and discussion

The variation of velocity 'u' with 'y' is calculated from equation (1.27) and is shown in figure (1.2) for different Hartmann number M with x=1, a=0.5, b=0.5, d=1.25, λ_1=0.5, ϕ=π/3, $\overline{Q}$=1 and β=0.01. It is observed that velocity profiles are parabolic for fixed values of the parameter 'M'. It is also observed that the velocity 'u' increases with increasing Hartmann number M near the walls. However, 'u' decreases by increasing 'M' near the centre of the channel.

The variation of velocity u with y is for different values of the average volume flow rate $\overline{Q}$ with x=1, a=0.5, b=0.5, d=1.25, M=1, λ_1=0.5, ϕ=π/3, and β=0.01 is shown in figure (1.3). It is concluded that the velocity increases with increasing the average volume flow rate $\overline{Q}$.

The relation between velocity u and y is drawn for different values of the Jeffrey material parameter λ_1 with x=1, a=0.5, b=0.5, d=1.25, M=1, ϕ=π/3, $\overline{Q}$=1 and β=0.01 is depicted in figure (1.4). It is noticed that velocity u increases with increasing Jeffrey material parameter λ_1 and act as an increasing resistant against the flow in the central part of the channel.

In figure (1.5) the variation of velocity is depicted for different values of phase difference ϕ with x=1, a=0.5, b=0.5, d=1.25, x =1, λ_1=0.5, M=1, $\overline{Q}$=1 and β=0.01. It is observed that increases in the phase difference ϕ decrease the velocity.

The relationship between u and y is plotted for different values of permeability parameter including slip β with a=0.5, b=0.5, d=1.25, λ_1=0.5, ϕ=π/3, $\overline{Q}$=1 and M=1 and is shown figure (1.6). It is found that the velocity u increases with increases in permeability parameter β near the walls and u decreases by increasing β near the centre of the channel.

The variation of pressure gradient $\frac{dp}{dx}$ with x is depicted in figure (1.7) to (1.11) for different values M and $\overline{Q}$. It is noticed that the magnitude of the pressure gradient $\frac{dp}{dx}$ increases with increasing Hartmann number M and average flow rate $|\overline{Q}|$. It is found that the pressure gradient $\frac{dp}{dx}$ decreases with increasing phase difference ϕ and the point of the maximum decreases with increasing ϕ. We also observed that increase in the Jeffrey material parameter λ_1 and permeability parameter β decrease the pressure gradient $\frac{dp}{dx}$.

The variation on pressure rise ΔP_λ with the average volume flow rate $\overline{Q}$ is calculated form equation (1.31) and is shown figure (1.12) for different values of Hartmann number M with a=0.5, b=0.5, d=1.25, $\phi= \pi/3$, $\lambda_1 = 0.5$ and $\beta=0.01$. We observed that for values of $\overline{Q}$ between 0.1 and 0.2, the pumping curves intersect at a point (0.12, 2). For a given flux $\overline{Q}$, the pressure rise ΔP_λ increases with decreasing Hartmann number M below this point and opposite behaviour is observed above this point.

The relation between pressure rise ΔP_λ with the average volume flow rate $\overline{Q}$ is depicted in figure (1.13) for different values of the Jeffrey material parameter λ_1 with a=0.5, b=0.5, d=1.25, M=1, $\phi= \pi/3$ and $\beta=0.01$. We observed that for values of $\overline{Q}$ between 0.4 and 0.5, the pumping curves intersect at a point (0.45, -1). For a given the average volume flow rate $\overline{Q}$, the pressure rise ΔP_λ increases with increasing the Jeffrey material parameter λ_1 below this point and opposite behaviour is observed above this point.

In figure (1.14) the relation between the average volume flow rate $\overline{Q}$ and pressure rise ΔP_λ is plotted at different values of the phase difference ϕ with a=0.5, b=0.5, d=1.25, M=1, $\phi= \pi/3$ and β=0.01. It is noticed that the pressure rise ΔP_λ decreases with increasing the phase difference ϕ.

The relation between the average volume flow rate $\overline{Q}$ and the pressure rise ΔP_λ is drawn for different values of the permeability parameter including slip β with a=0.5, b=0.5, d=1.25, M=1, $\phi=\pi/3$ and $\lambda_1=0.5$ is depicted in figure (1.15). We observed that for values of $\overline{Q}$ between 0.4 and 0.5, the pumping curves intersect at a point (1.43, -0.7). For a given flux $\overline{Q}$, the pressure rise ΔP_λ increases with increasing the permeability parameter β below this point and opposite behaviour is observed above this point.

The axial shear stress distribution S_{xy} on the upper wall of an asymmetric channel for different values of Hartmann number with a=0.5, b=0.5, d=1, M=1, ϕ $=\pi/6$, $\lambda_1=0.5$, and β =0.1. We noticed that the stress is in oscillatory behaviour with increasing Hartmann number M.

The relation between the axial shear stress S_{xy} on the upper wall with y is depicted in figure (1.24) for different values of the average volume flow rate $\overline{Q}$ with a=0.5, b=0.5, d=1, $\lambda_1=0.5$, M=2, $\phi=\pi/6$ and β=0.1. We observed that the curves intersect at a point (0, 0), shear stress S_{xy} increases with increasing the average volume flow rate $\overline{Q}$ below this point and opposite behaviour is observed above this point.

The relation between the axial shear stress S_{xy} on the upper wall with y is depicted in figure (1.25) for different values of the phase difference ϕ with a=0.5, b=0.5, d=1, $\lambda_1=0.5$, M=2, $\overline{Q}=2$ and β=0.1. We observed that the shear stress S_{xy} increases with increasing the phase difference.

The variation of the axial shear stress S_{xy} on the upper wall with y is plotted in figure (1.26) for different values the permeability parameter β with a=0.5, b=0.5, d=1, M=2, λ_1=0.5, $\overline{Q}$=2, ϕ=π/6 and β=0.1. We observed that the curves intersect at a point (0, 0), shear stress S_{xy} increases with increasing the permeability parameter β below this point and opposite behaviour is observed above this point.

The relation between the axial shear stress on the upper wall S_{xy} with y is plotted in figure (1.27) for different values the Jeffrey material parameter λ_1 with a=0.5, b=0.5, d=1, M=2, $\overline{Q}$=2, λ_1=0.5, ϕ=π/6 and β=0.1. We observed that the curves intersect at a point (0, 0), shear stress S_{xy} increases with increasing the Jeffrey material parameter λ_1 below this point and opposite behaviour is observed above this point.

1.6.1 Trapping phenomena

The phenomenon of trapping is another interesting topic in peristalsis. The formation of an internally circulating bolus of the fluid by closed streamlines is called trapping and this trapped bolus will be pushed ahead along the peristaltic wave. The effects of permeability parameter β, and average flow rate $\overline{Q}$, on the trapping are illustrated in figures (1.28) and (1.32). It observed that the bolus increases with increasing β and $\overline{Q}$ and disappears for $\overline{Q}$=1.4. The effects of material parameter λ_1, Hartmann number M, and channel width d on the trapping for different wave forms can be seen through figures (1.29), (1.30) and (1.35) for fixed values of other parameters. It can be concluded that increases in λ_1, M, and d result in the decrease in the volume of trapping bolus. Moreover bolus are disappear for λ_1 =3, M=3 and d=1.6. When the amplitudes either a or b equal to zero, peristaltic wave exists only on one of walls, the mechanism of trapping remains the same the trapping phenomenon at the central line and near the boundary of the channel are shown in figures (1.33) and (1.34).

The effects of phase difference ϕ on trapping with same amplitudes a=b=0.5 for $\overline{Q}$=2 is plotted in figure (1.31). It is observed that the bolus appearing in the central region for $\phi=0$ moves towards left and decreases in size as ϕ increases. For ϕ $= \pi$ the bolus disappears and stream lines are parallel to the boundary walls.

Stream lines are plotted in figures (1.36) and (1.37) for different wave forms, namely a) sinusoidal b) triangular c) trapezoidal wave for various β. We find that the size of trapped bolus decreases for all these three wave form with an increasing β. We noticed that the size of the bolus in greater for trapezoidal wave when compared with sinusoidal and triangular waves for given β.

The influence of Jeffrey parameter λ_1 on the trapped bolus is analyzed and is plotted in figures (1.38) and (1.39). We observed that the for $\lambda_1=0$, there is no trapping but for $\lambda_1=1$ trapped bolus is observed for all three waves forms considered for given $\lambda_1(\neq 0)$. We noticed that the size of the trapped bolus is higher for trapezoidal wave forms and less in case of sinusoidal wave form.

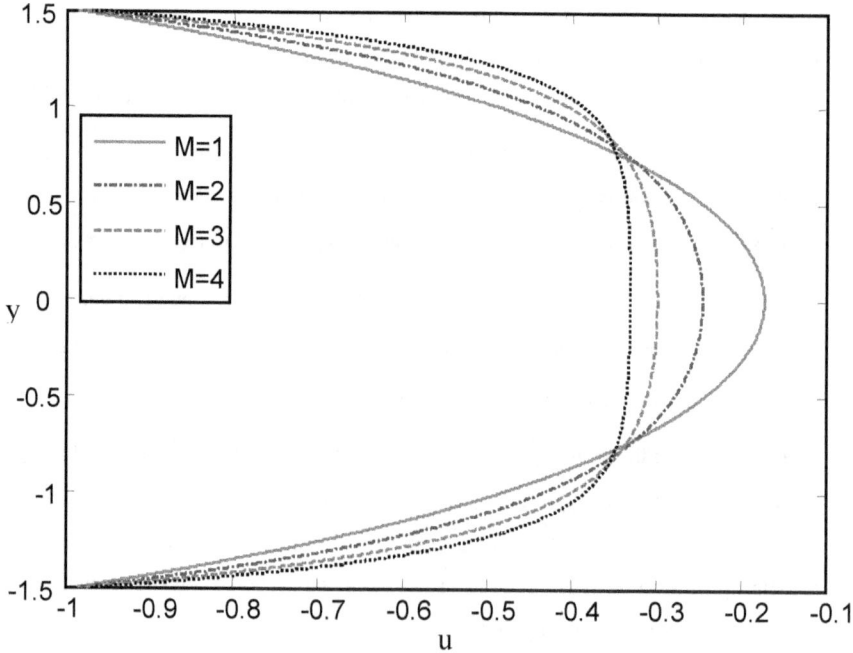

Figure 1.2: The velocity profiles with a=0.5, b=0.5, d=1.25 x=1, $\phi=\pi/3$, $\overline{Q}=1$, λ_1 =0.5 and β=0.01.

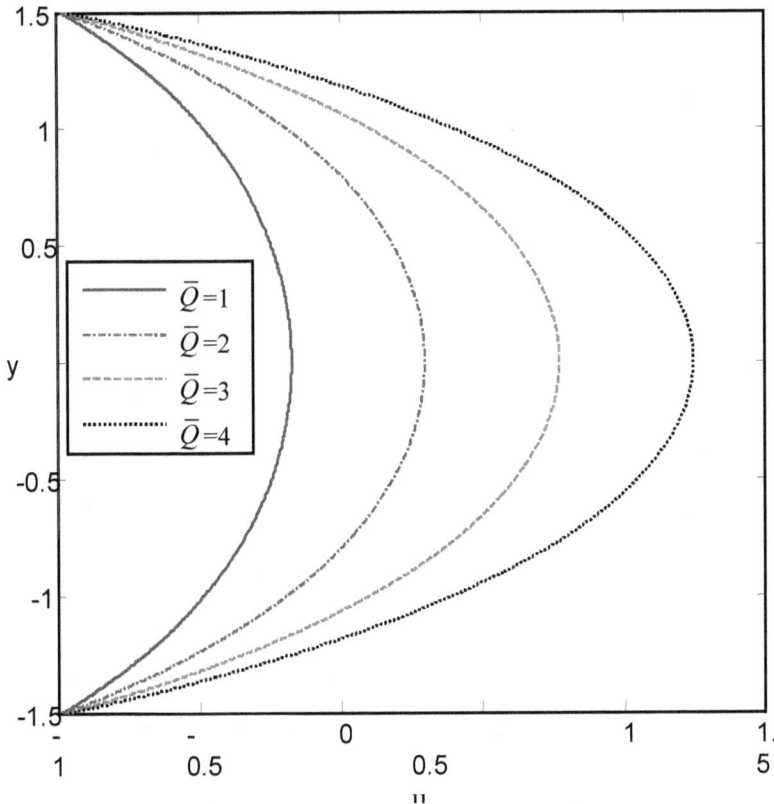

Figure 1.3: The velocity profiles with a=0.5, b=0.5, d=1.25 x=1, M=1, $\phi=\pi/3$,

λ_1 =0.5 and β=0.01.

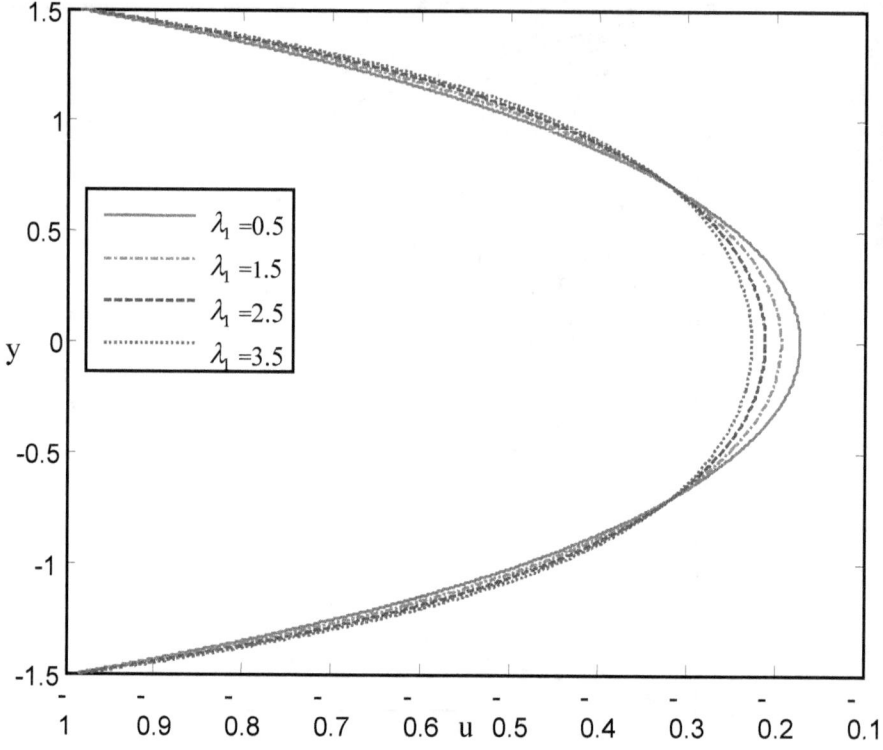

Figure 1.4: The velocity profiles with a=0.5, b=0.5, d=1.25 x=1, M=1, $\phi=\pi/3$, $\overline{Q}$=1 and β=0.01.

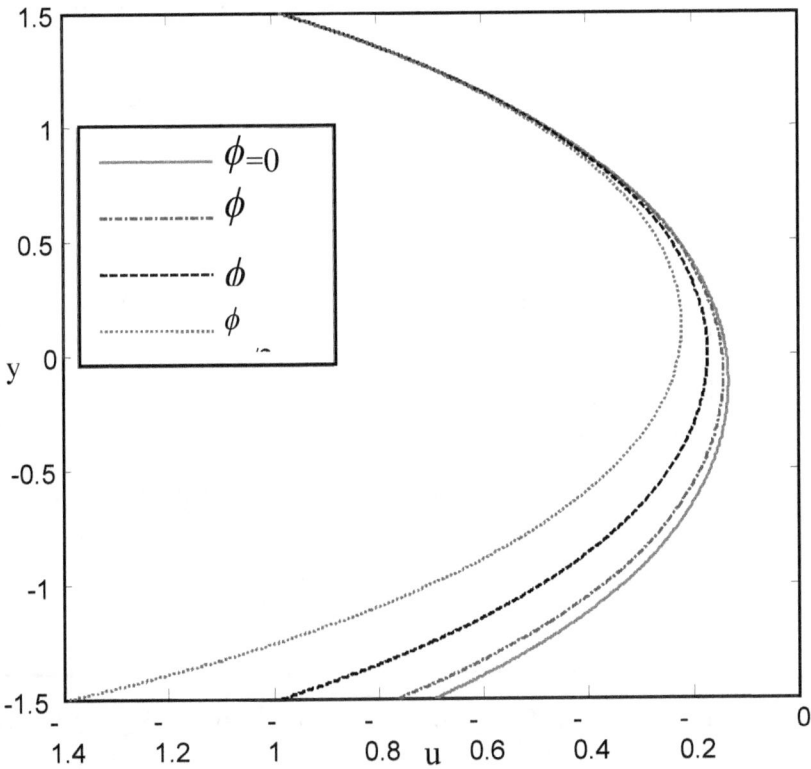

Figure 1.5: The velocity profiles with a=0.5, b=0.5, d=1.25 x=1, M=1, $\overline{Q}$=1, λ_1 =0.5 and β=0.01.

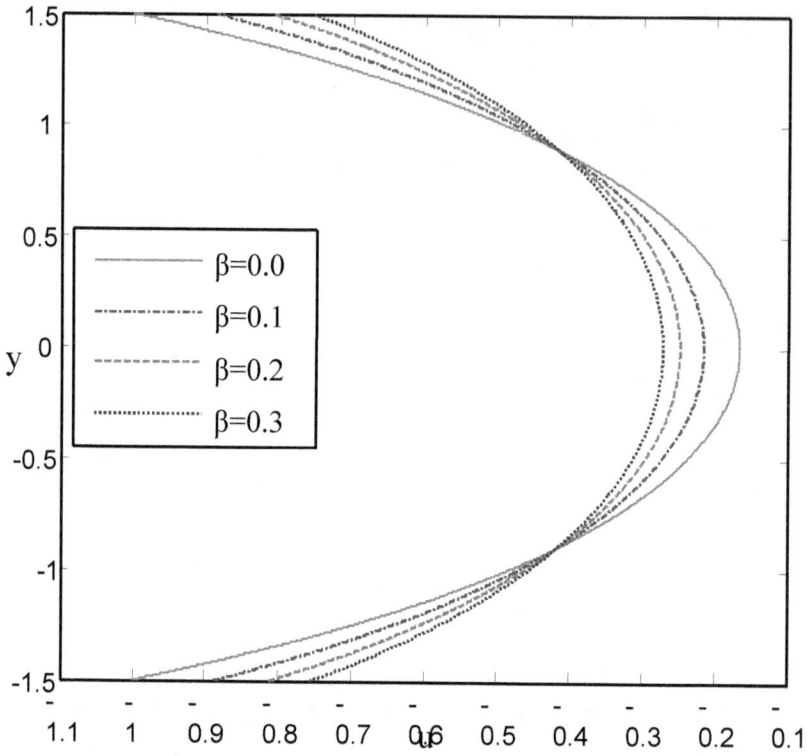

Figure 1.6: The velocity profiles with a=0.5, b=0.5, d=1.25 x=1, M=1, $\phi=\pi/3$, $\overline{Q}=1$ and λ_1 =0.5.

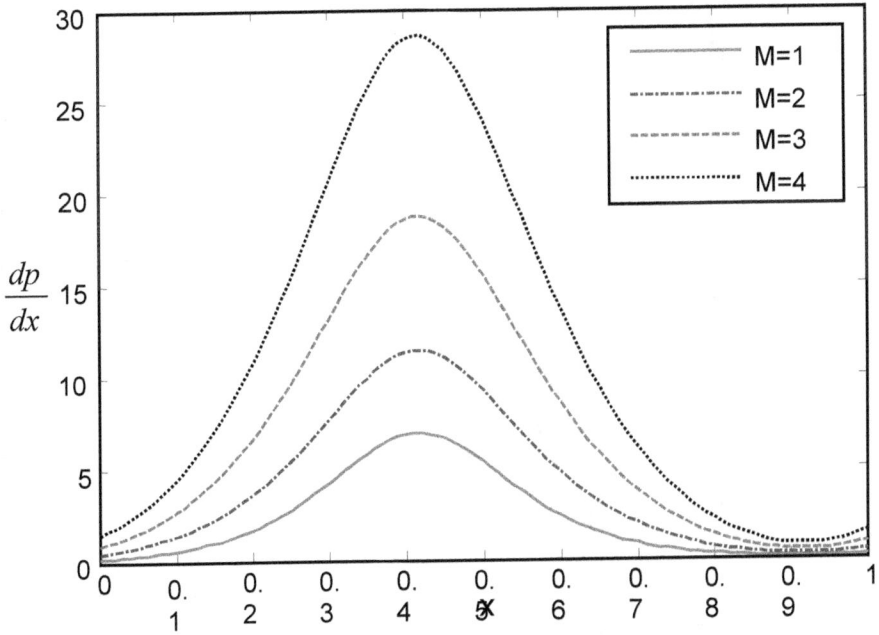

Figure 1.7: Variation of pressure gradient $\dfrac{dp}{dx}$ with x for different values of

Hartmann number M with a=0.5, b=0.5, d=1.25, ϕ=π/3, $\overline{Q}$=-1,

λ_1 =0.5 and β=0.01;

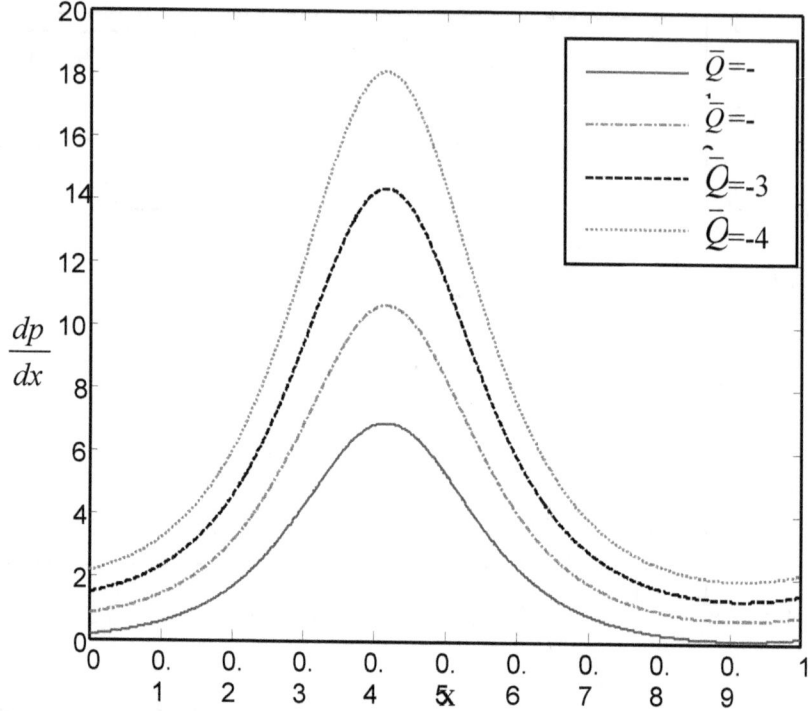

Figure 1.8: Variation of pressure gradient $\frac{dp}{dx}$ with x for different values of   mean

flow rate $\overline{Q}$ with a=0.5, b=0.5, d=1.25, M=1, $\phi=\pi/3$, λ_1 =0.5 and

β=0.01.

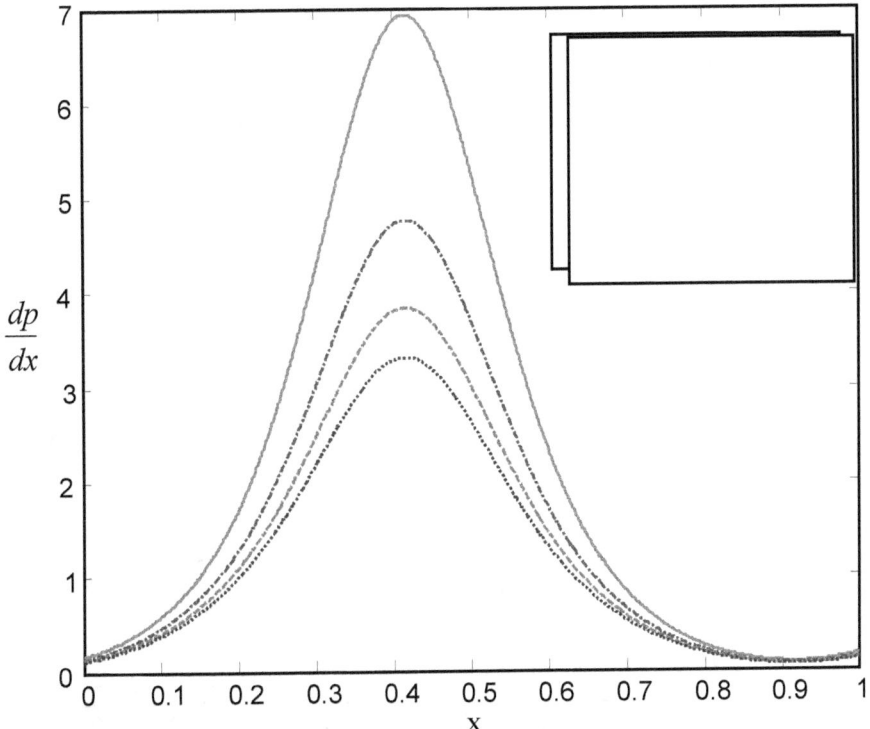

Figure 1.9: Variation of pressure gradient $\frac{dp}{dx}$ with x for different values of Jeffrey

material parameter λ_1 with a=0.5, b=0.5, d=1.25, M=1, ϕ=π/3, $\overline{Q}$=-1

and β=0.01.

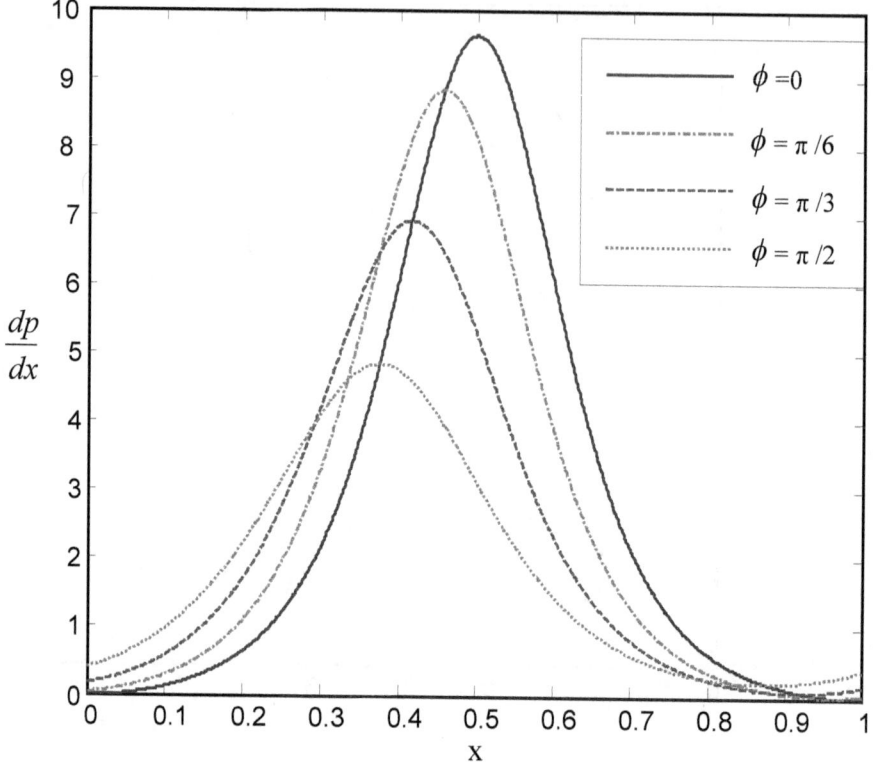

Figure 1.10: Variation of pressure gradient $\dfrac{dp}{dx}$ with x for different values of phase

difference ϕ with a=0.5, b=0.5, d=1.25, M=1, $\overline{Q}$=-1, $\quad \lambda_1$ =0.5

and β=0.01.

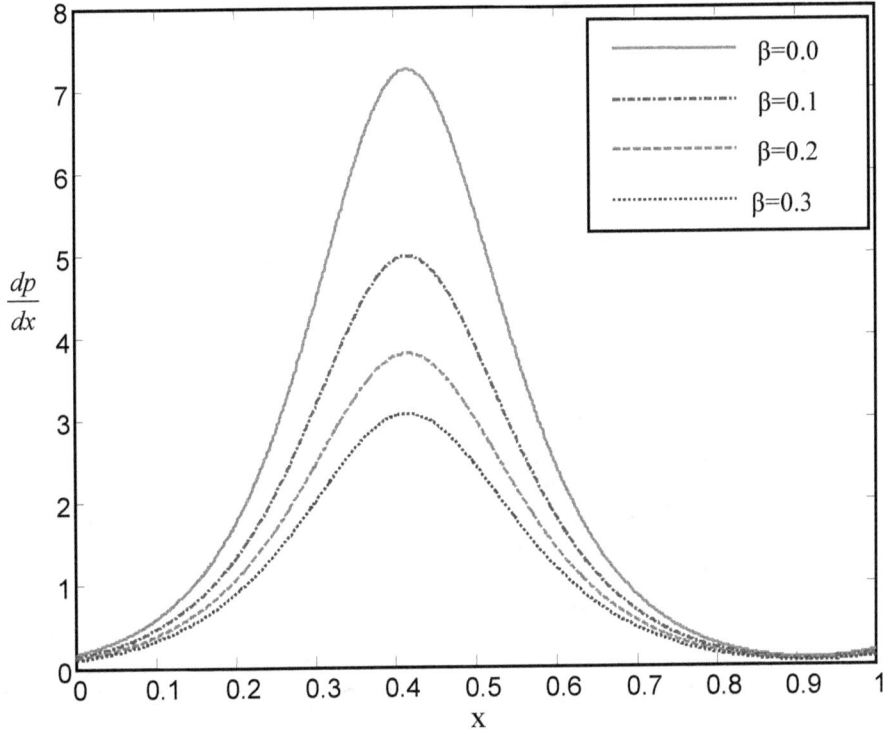

Figure 1.11: Variation of pressure gradient $\frac{dp}{dx}$ with x for different values of permeability parameter including slip β with a=0.5, b=0.5, d=1.25; M=1, $\phi=\pi/3$; $\overline{Q}$=-1 and λ_1 =0.5;

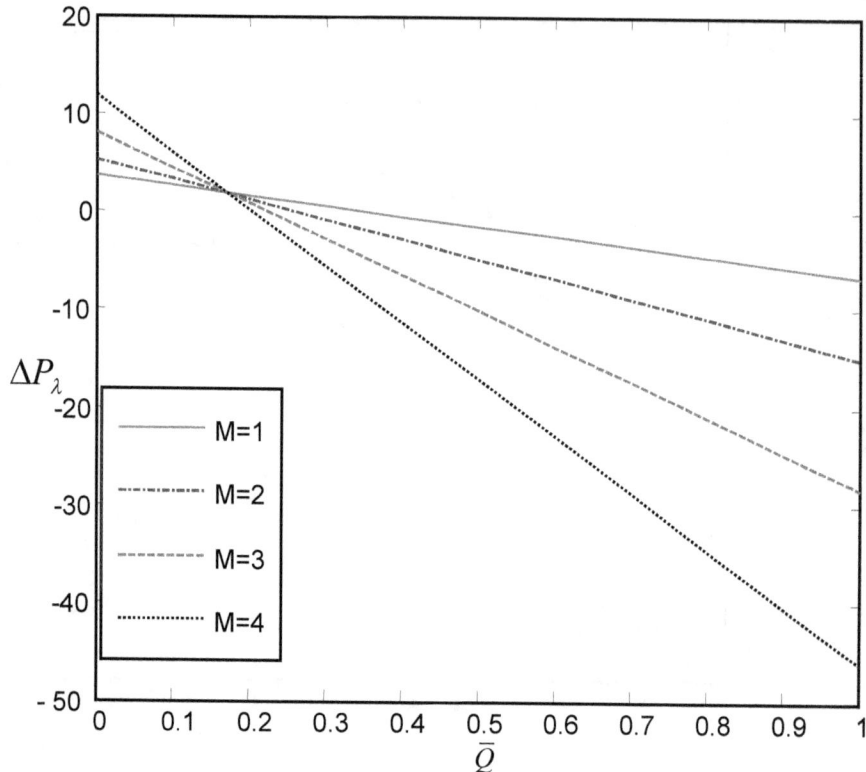

Figure 1.12 Variation of $\bar{Q}$ with ΔP_λ for different value of Hartmann number M with a=0.5, b=0.5, d=1.25, $\phi= \pi/3$, $\lambda_1 =0.5$ and β=0.01;

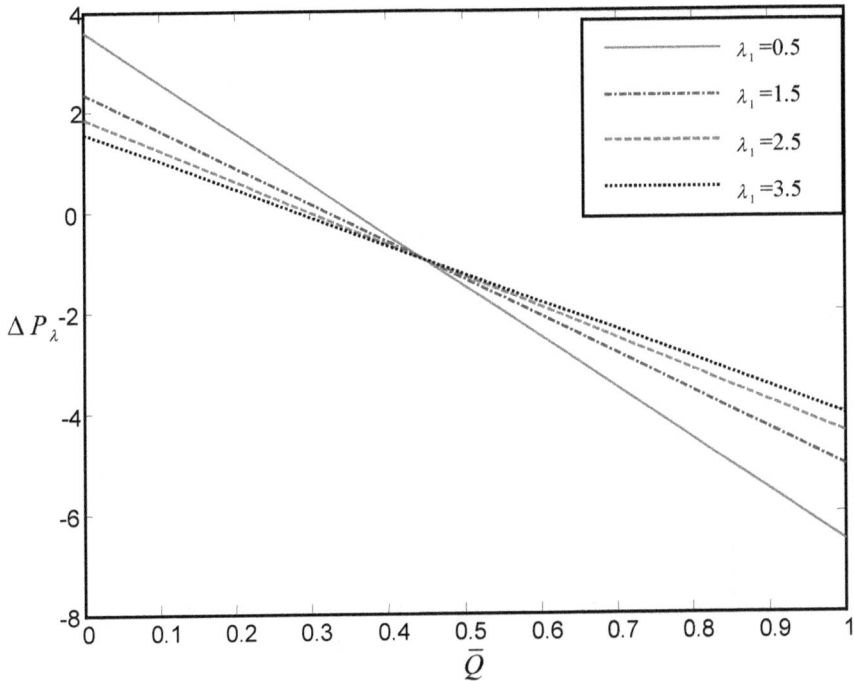

Figure 1.13: Variation of $\bar{Q}$ with ΔP_λ for different values of Jeffrey material parameter λ_1 with a=0.5, b=0.5, d=1.25, M=1, $\phi= \pi/3$ and β=0.01.

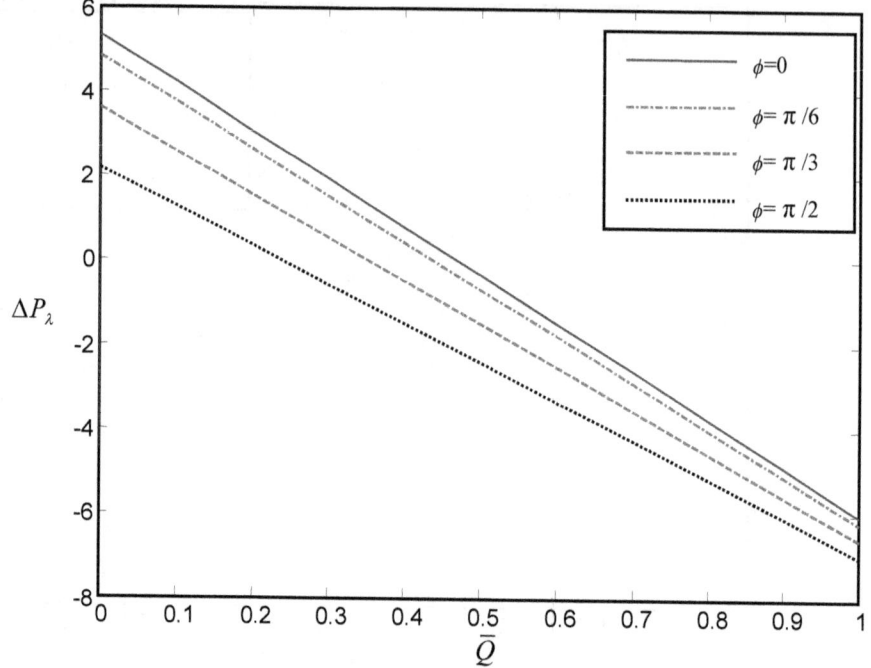

Figure 1.14: Variation of $\overline{Q}$ with ΔP_λ for different value of phase difference ϕ with a=0.5, b=0.5, d=1.25, M=1, λ_1=0.5 and β=0.01.

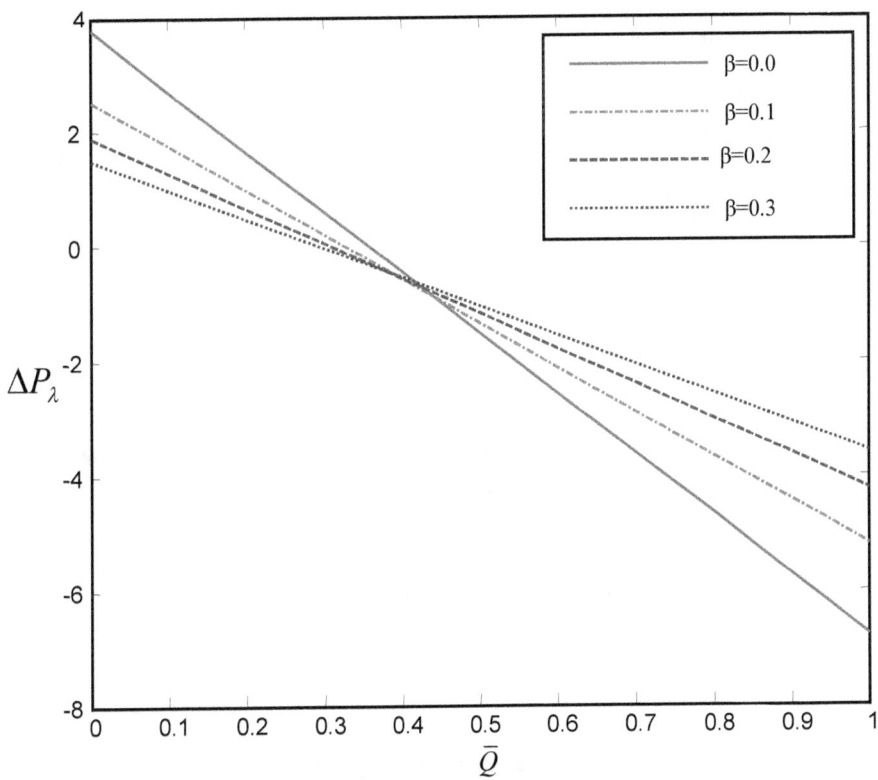

Figure 1.15: Variation of $\overline{Q}$ with ΔP_λ for different of permeability parameter including slip β with a=0.5, b=0.5, d=1.25, M=1, $\phi= \pi /3$ and λ_1 =0.5;

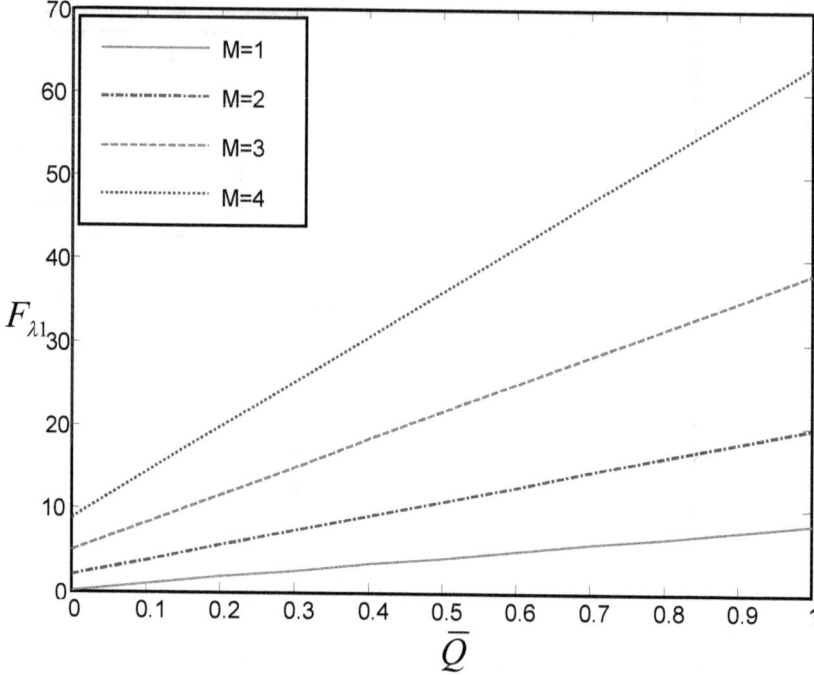

Figure 1.16: Variations of frictional force $F_{\lambda 1}$ with y=h$_1$ for a=0.5, b=0.5, d=1.25, ϕ = π /3 and λ_1 =0.5, β=0.01.

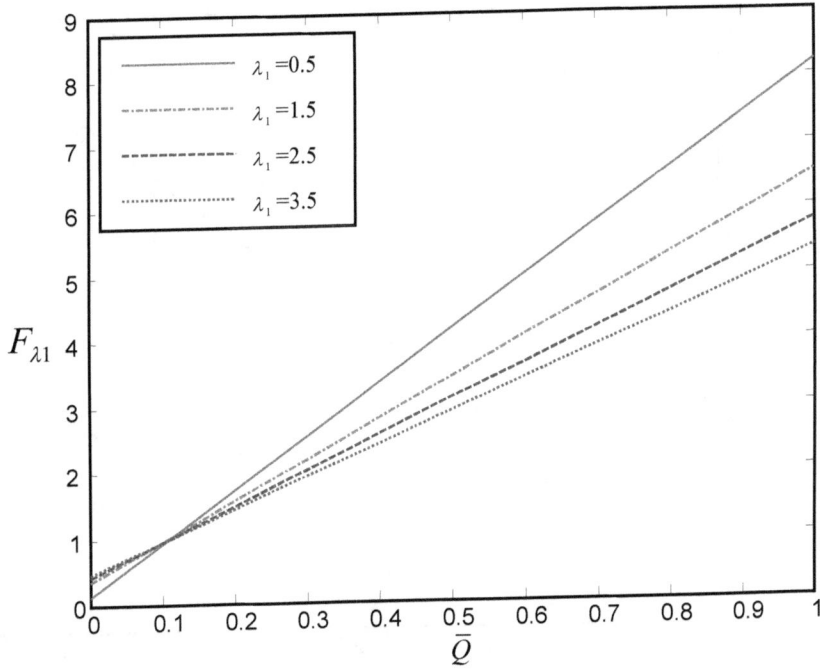

Figure 1.17: Variations of frictional force F_{λ_1} at y=h₁ for a=0.5, b=0.5, d=1.25, $\phi= \pi$ /3, M=1 and β=0.01.

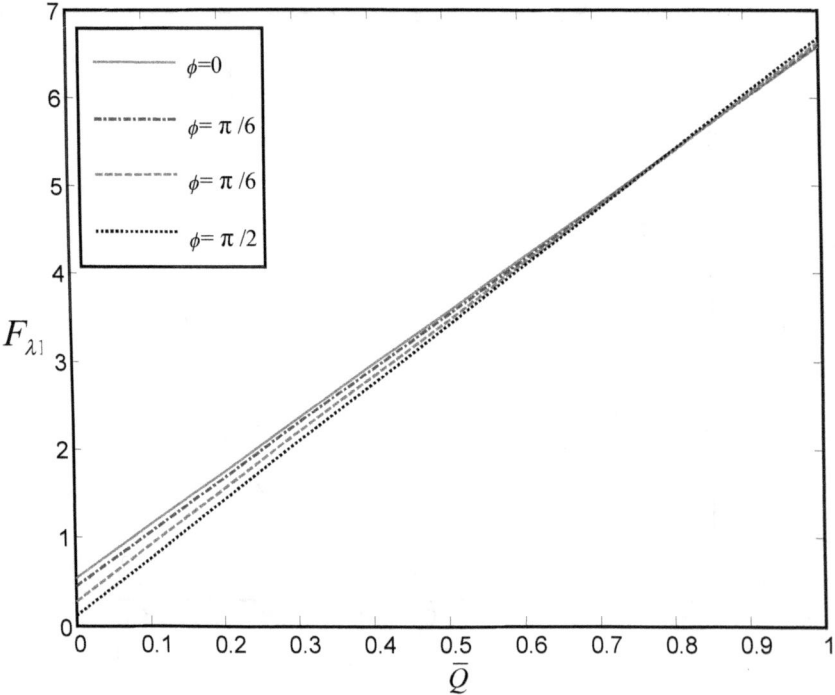

Figure 1.18: Variations of frictional force $F_{\lambda 1}$ at y=h$_1$ for a=0.5, b=0.5, d=1.25, λ_1 =0.5, M=1 and β=0.01.

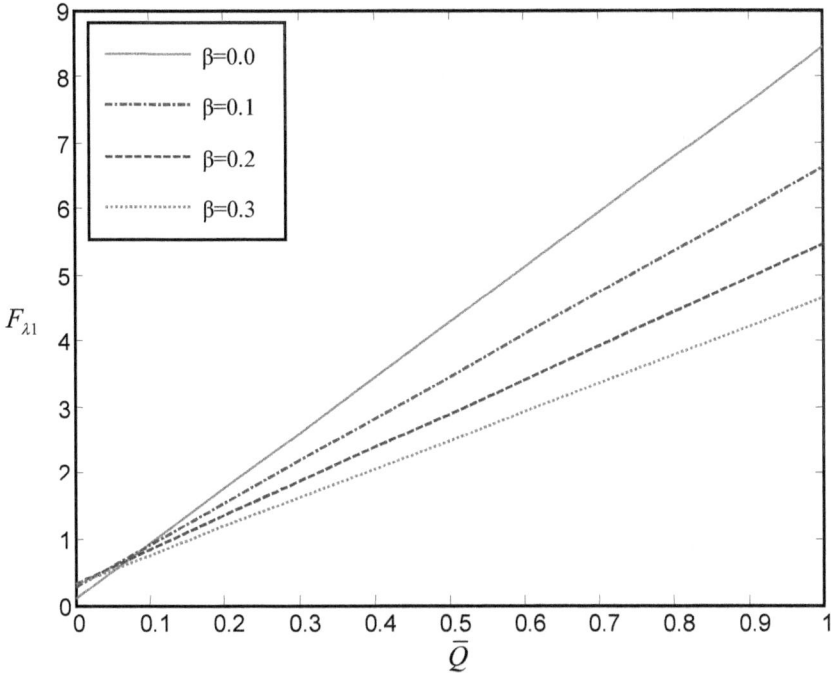

Figure 1.19: Variations of frictional force $F_{\lambda 1}$ at y=h$_1$ for a=0.5, b=0.5, d=1.25,

$\phi=\pi/3$, λ_1 =0.5 and M=1.

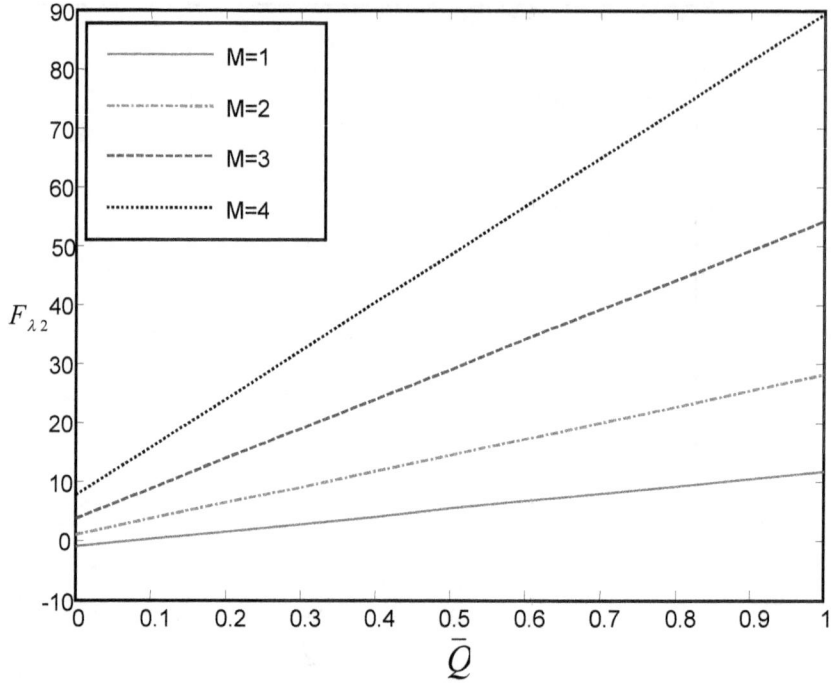

Figure 1.20: Variations of frictional force F_{λ_2} at y=h₂ for a=0.5, b=0.5, d=1.25, ϕ= π/3, λ_1=0.5 and β=0.01.

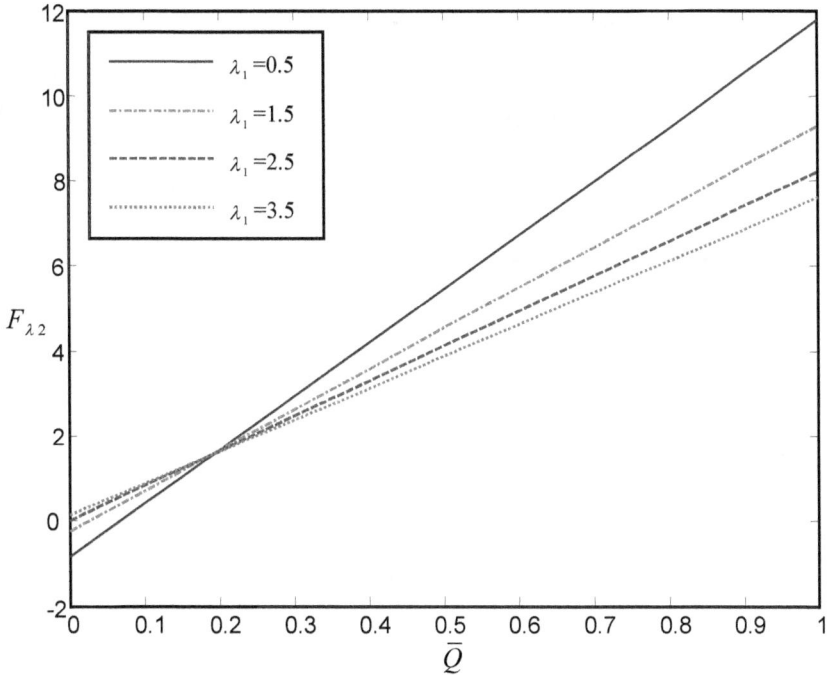

Figure 1.21: Variations of frictional force $F_{\lambda 2}$ at y=h$_2$ for   a=0.5, b=0.5,d=1.25, ϕ= π /3, M=1and β=0.01.

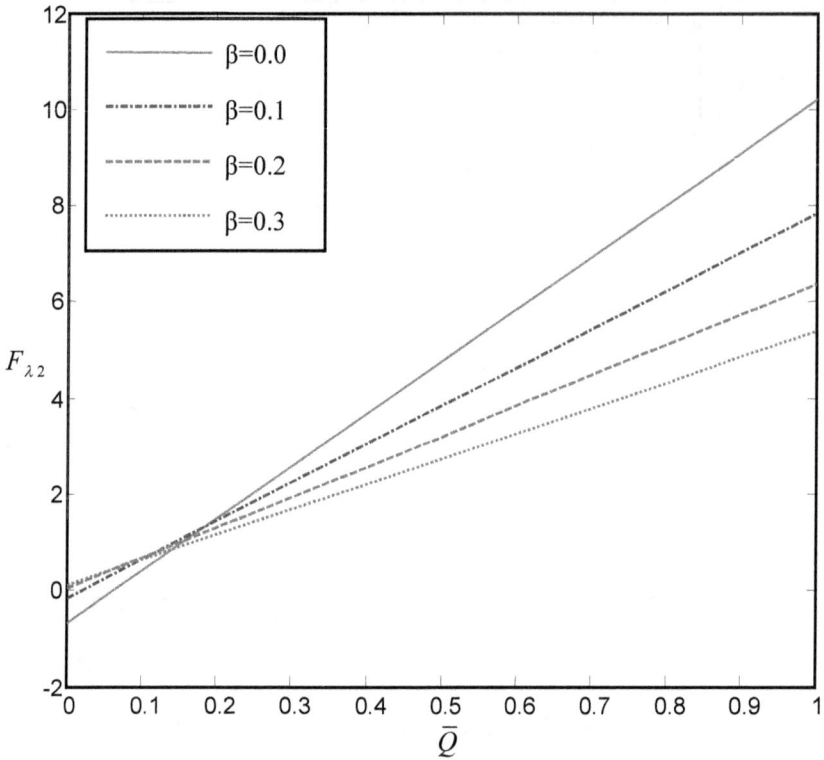

Figure 1.22: Variations of frictional force $F_{\lambda 2}$ at y=h$_2$ for a=0.5, b=0.5, d=1.25, ϕ= π/3, λ_1 =0.5 and M=1.

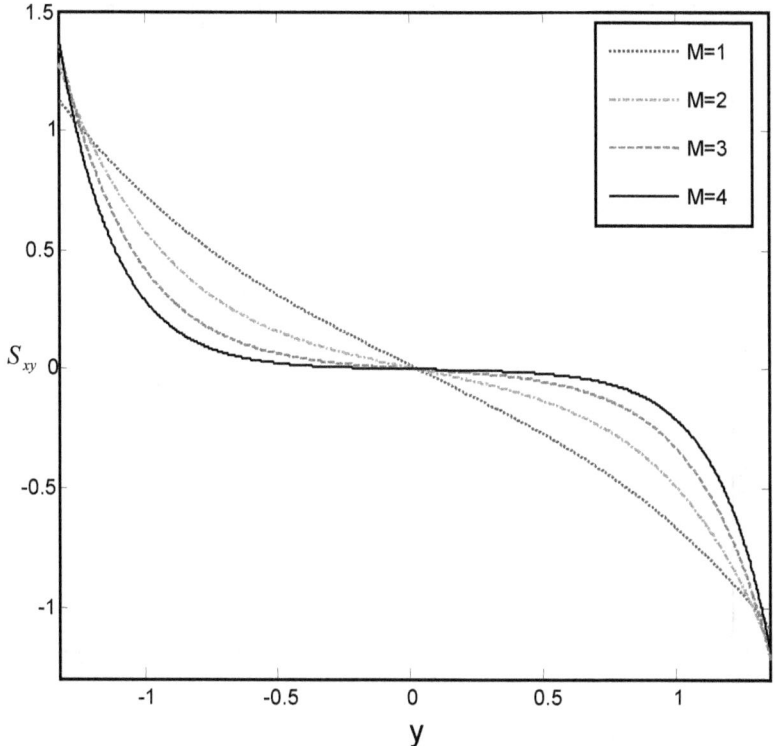

Figure 1.23: The axial shear stress distributions at the upper wall with a=0.5; b=0.5, d=1, $\phi = \pi/6$, $\overline{Q}$=2, λ_1 =0.5 and β=0.1.

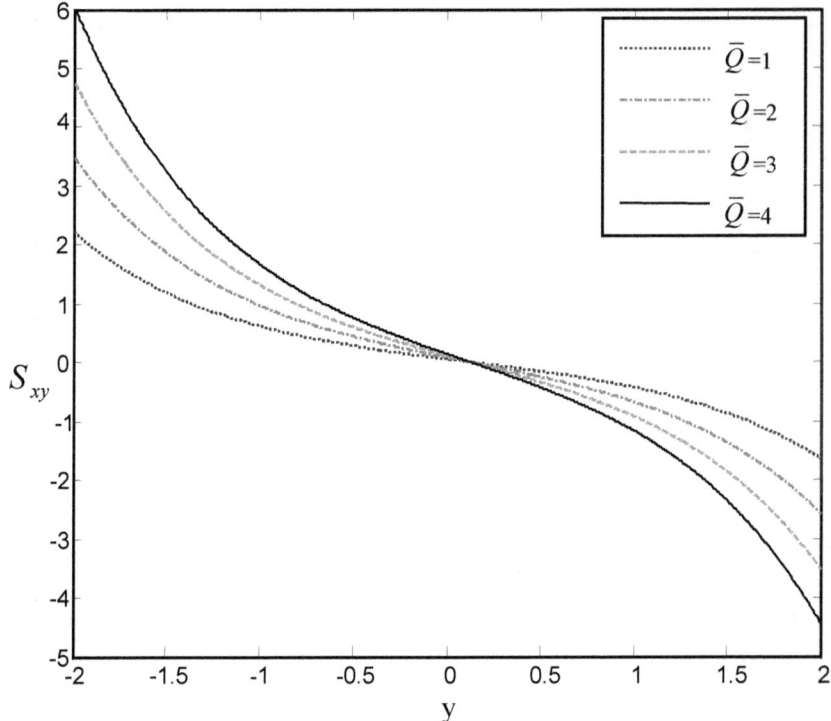

Figure 1.24: The axial shear stress distributions at the upper wall with a=0.5, b=0.5,

d=1, $\phi = \pi/6$, λ_1 =0.5, β=0.1and M=2.

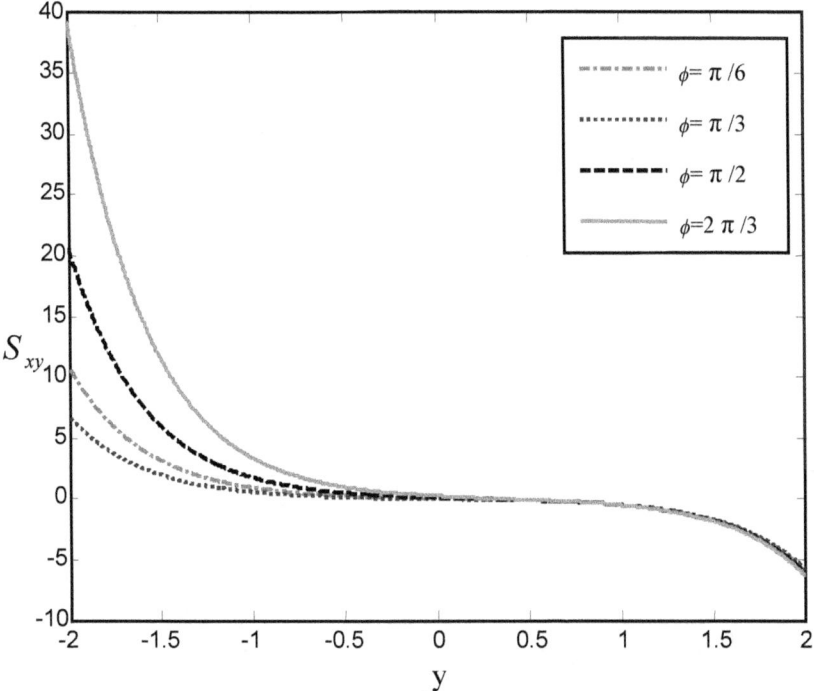

Figure 1.25: The axial shear stress distributions at the upper wall with

a =0.5, b=0.5, d=1, $\bar{Q}$=2, λ_1 =0.5, β=0.1 and M=2.

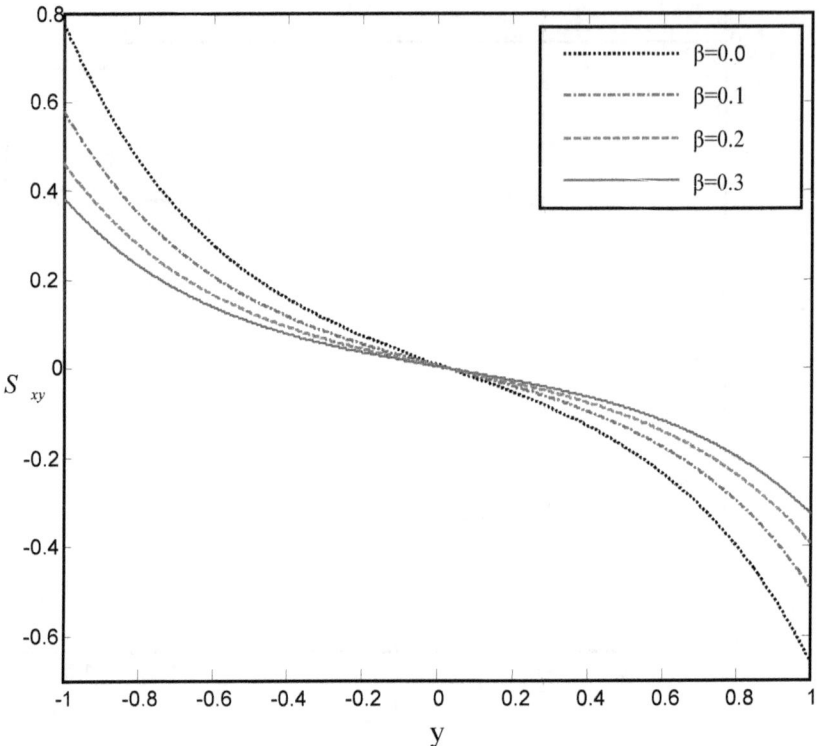

Figure 1.26: The axial shear stress distributions at the upper wall with a=0.5, b=0.5,

d=1, $\phi = \pi/6$, $\overline{Q}$=2, λ_1 =0.5 and M=2.

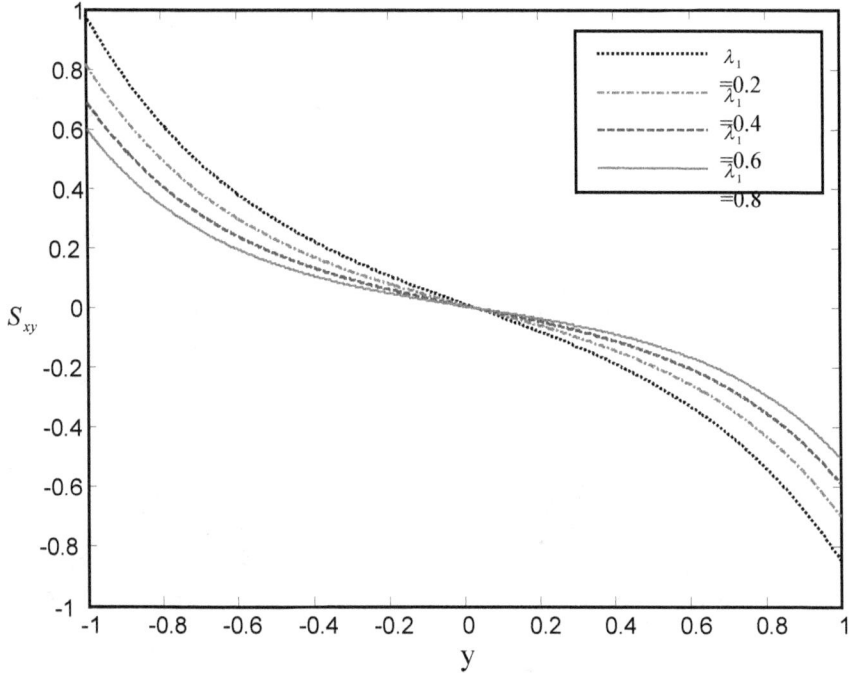

Figure 1.27: The axial shear stress distributions at the upper wall with a= 0.5, b=0.5,

d=1, $\phi=\pi/6$, $\overline{Q}$ =2, M=2 and β=0.1.

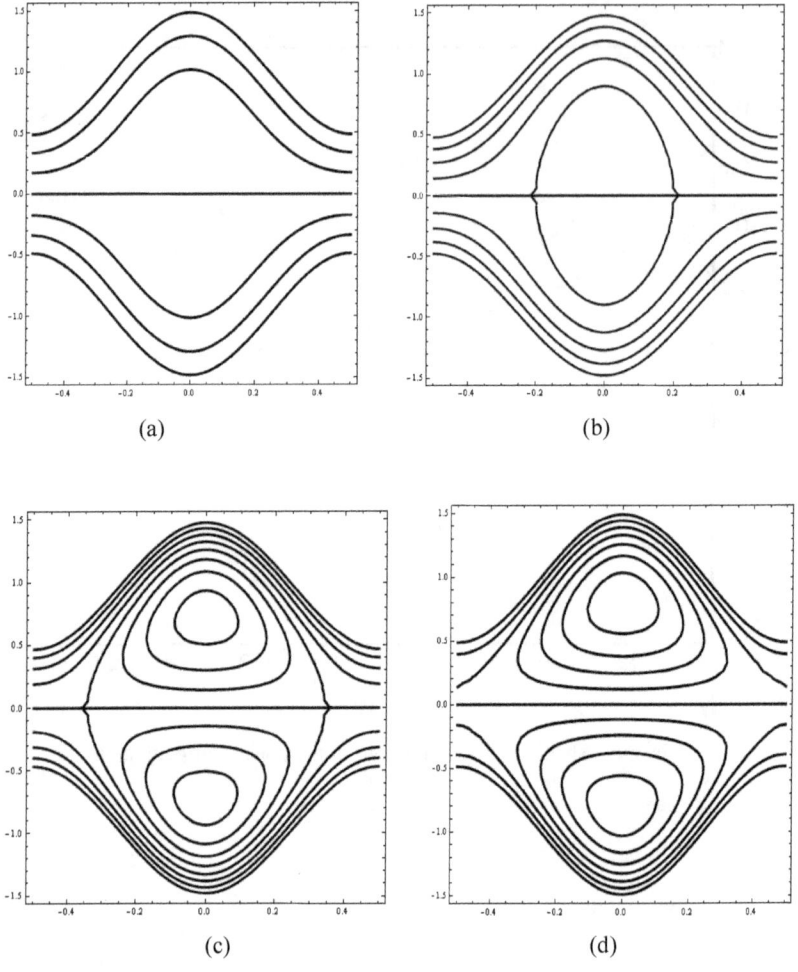

Figure 1.28: Streamlines in the wave frame for pumping with a=0.5, b=0.5, d=1, M=0.5, ϕ=0, λ_1=0.1, β=0.5, (a) $\overline{Q}$=1.4, (b) $\overline{Q}$=1.6, (c) $\overline{Q}$=1.6 and (d) $\overline{Q}$=1.7.

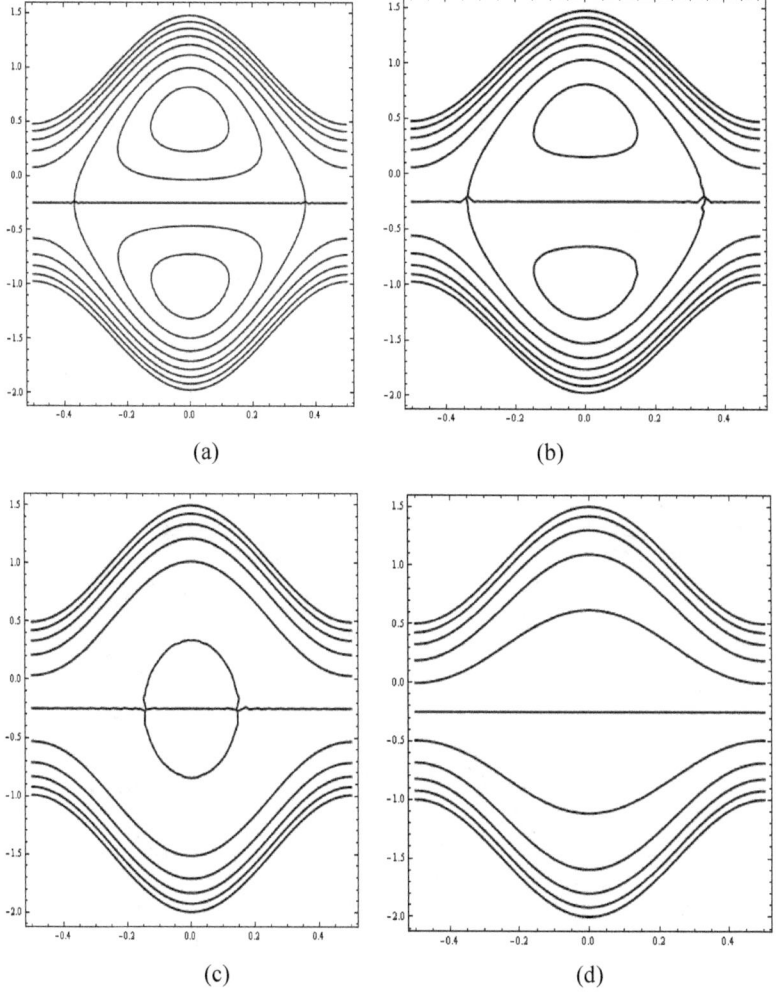

Figure 1.29: Streamlines in the wave frame for pumping with a=0.5, b=0.5, d=1.5, ϕ =0; λ_1=1, β =0.5, $\overline{Q}$=2, (a) M=0.1, (b) M=1, (c) M=2 and (d) M=3.

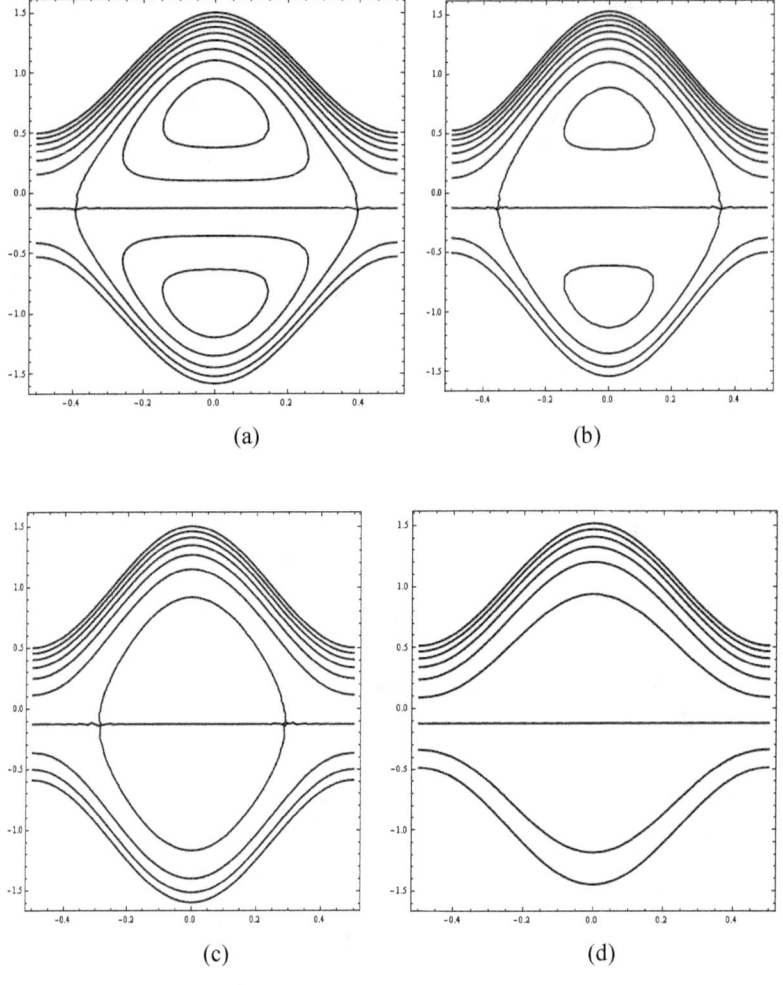

Figure 1.30: Streamlines in the wave frame for pumping with a=0.5, b=0.5, d=1.25, ϕ=0, β =0.25, $\overline{Q}$ =2, M=2, (a) λ_1 =0.1, (b) λ_1 =1, (c) λ_1 =2 and (d) λ_1 =3.

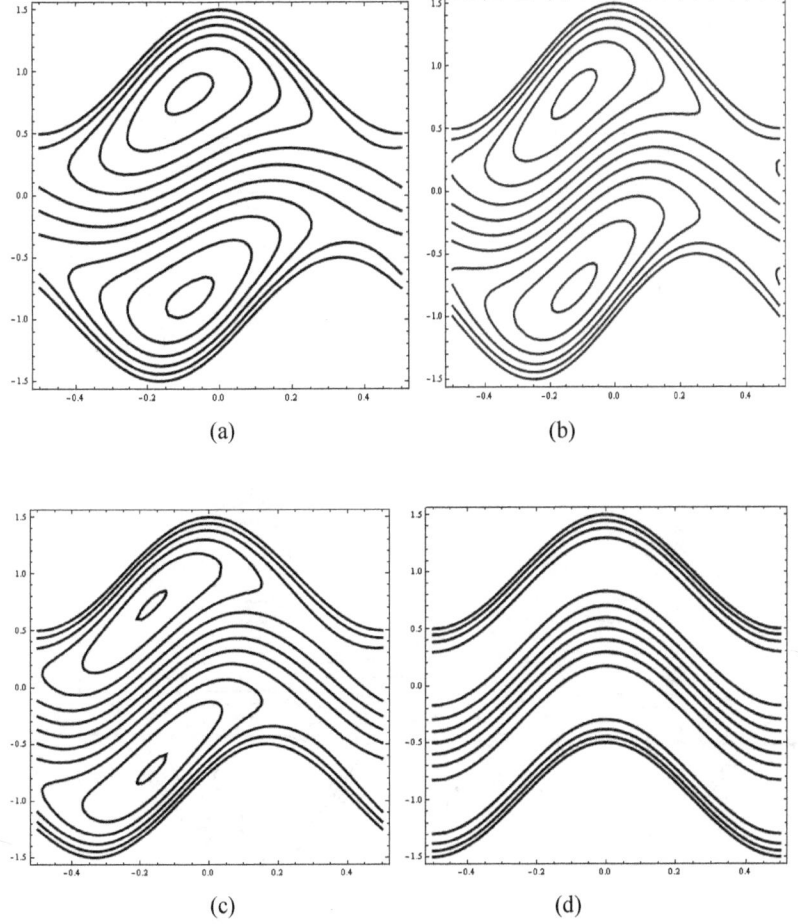

Figure 1.31: Streamlines in the wave frame for pumping with a=0.5, b=0.5, d=1, λ_1 =0.1, β =0.5, $\overline{Q}$ =2, M=0.5, (a) Φ = $\pi/3$, (b) ϕ = $\pi/2$, (c) ϕ = $2\pi/3$ and (d) ϕ = π .

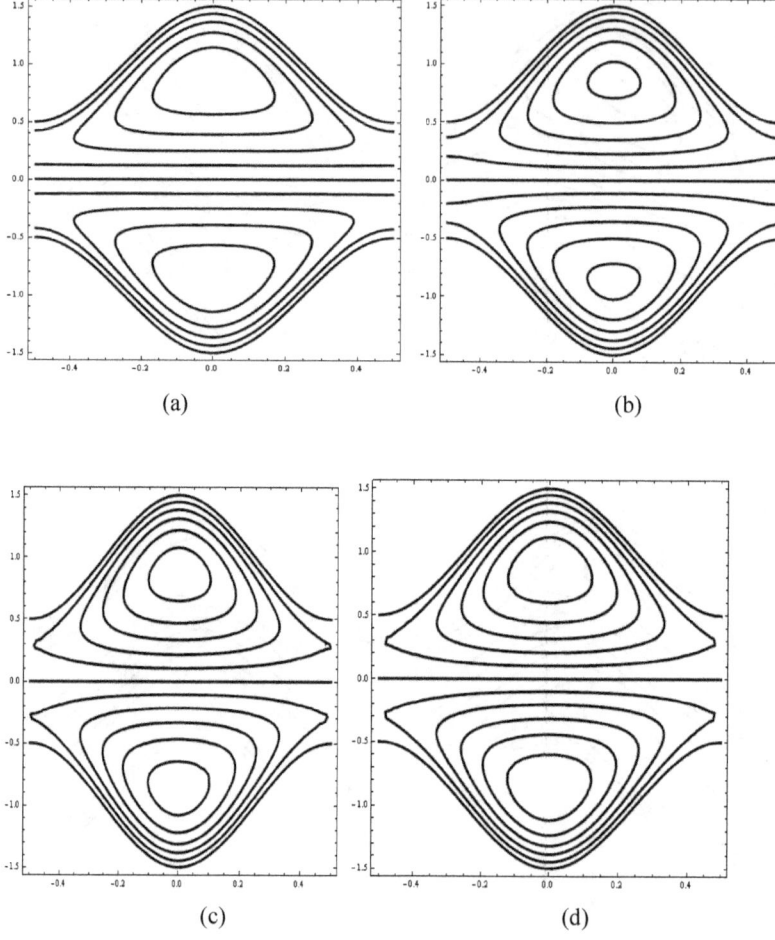

Figure 1.32: Streamlines in the wave frame for pumping with a=0.5, b=0.5, d=1, ϕ =0; $\overline{Q}$=2, λ_1 =0.1, M=0.5, (a) β =0.0, (b) β =0.2, (c) β =0.4 and (d) β =0.6.

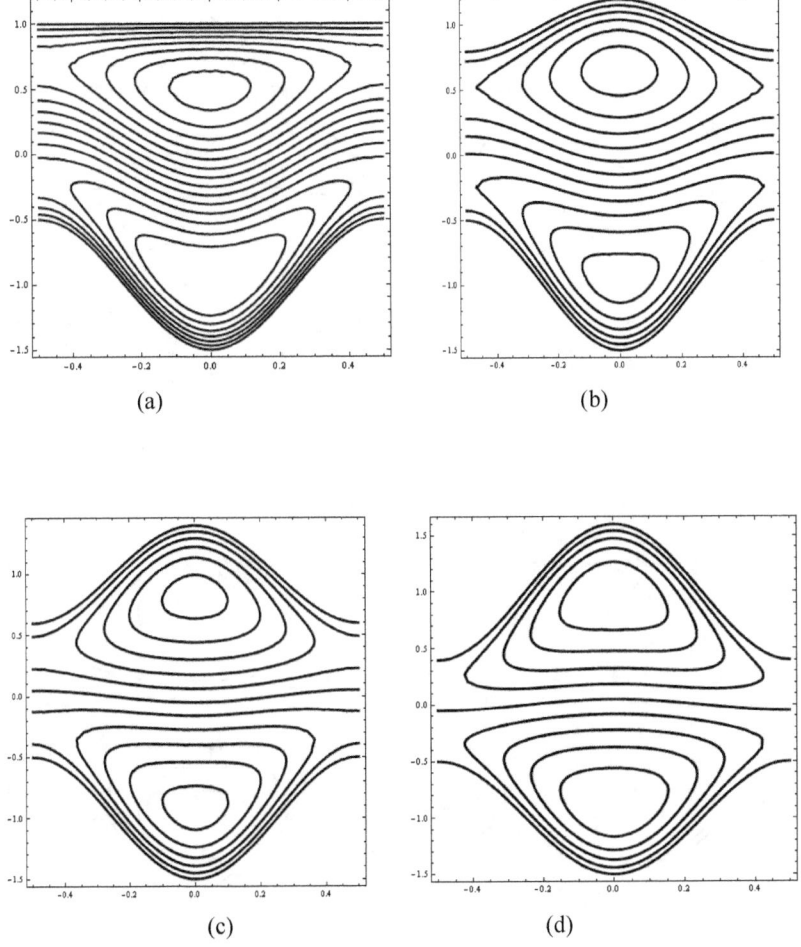

Figure 1.33: Streamlines in the wave frame for pumping with b=0.5, d=1, ϕ=0; $\overline{Q}$ =2, M=1, λ_1 =0.5, β =0.5, (a) a=0.0, (b) a=0.2, (c) a=0.4 and (d) a=0.6.

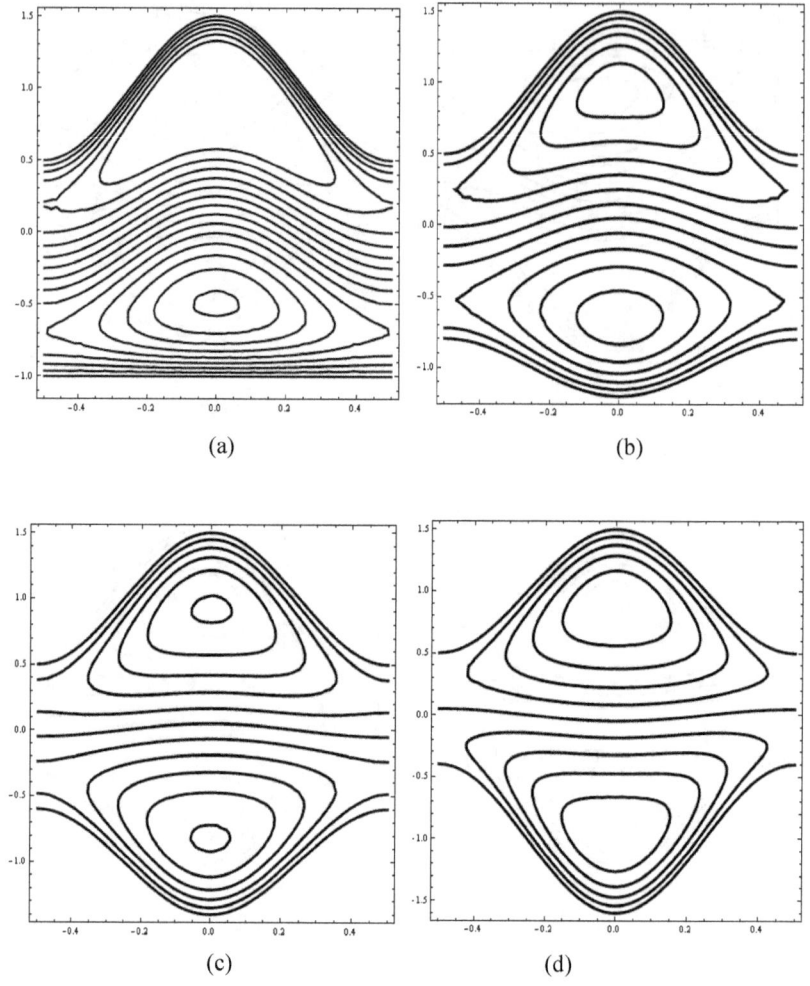

Figure 1.34: Streamlines in the wave frame for pumping with a=0.5, d=1, ϕ=0, $\overline{Q}$ =2, M=1, λ_1 =0.5, β =0.5, (a) b=0.0, (b) b=0.2, (c) b=0.4 and (d) b=0.6.

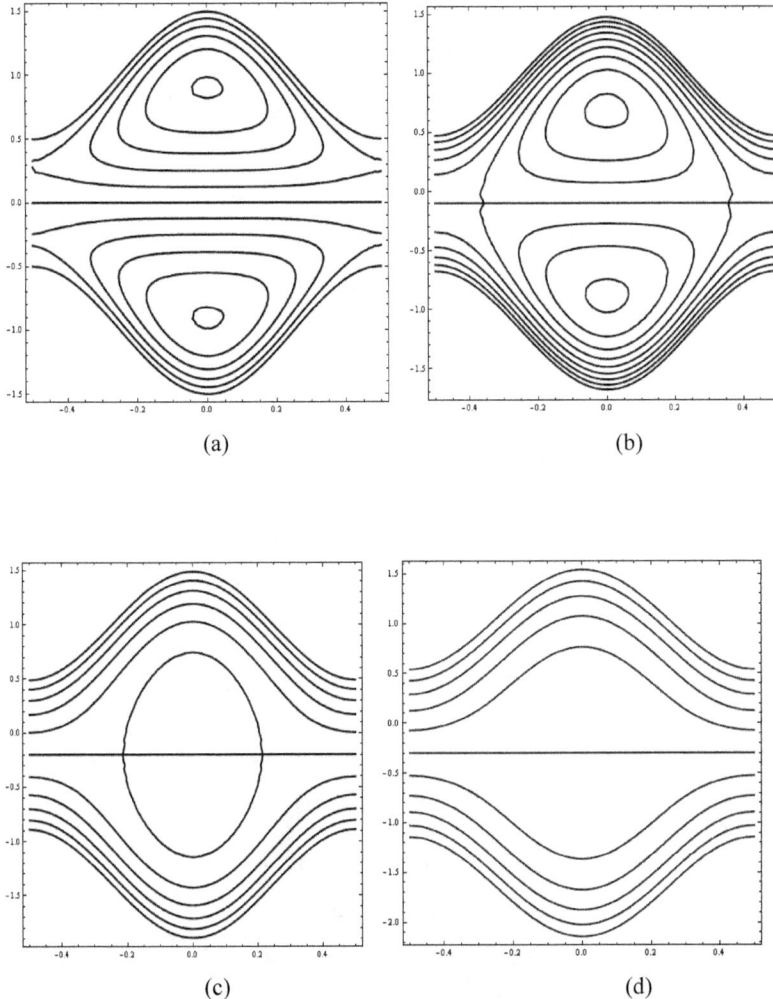

Figure 1.35: Streamlines in the wave frame for pumping with a=0.5, b=0.5, ϕ=0; $\overline{Q}$ =2, M=1, λ_1=1, β =0.5, (a) d=1.0, (b) d=1.2, (c) d=1.4 and (d) d=1.6.

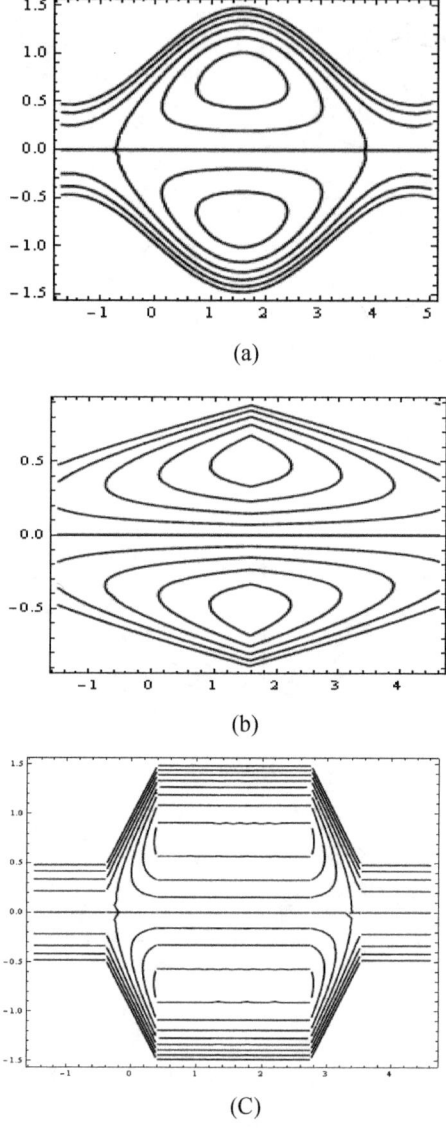

(a)

(b)

(C)

Figure 1.36: Streamlines for a=0.5, b=0.5, d=1, ϕ=0; M=1, $\bar{\zeta}$=1.8; λ_1=0.1, β=0.4, a) Sinusoidal wave, b) Triangular wave and c) Trapezoidal wave.

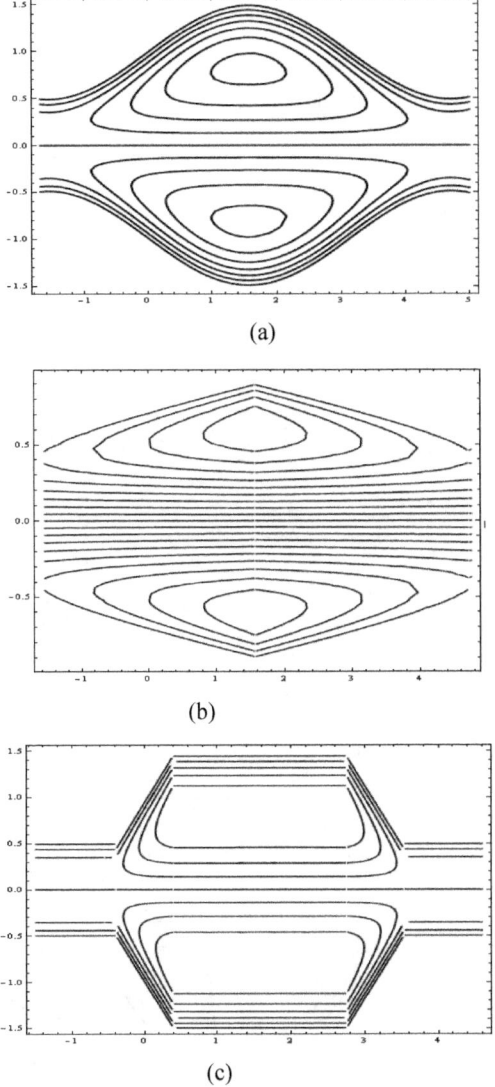

Figure 1.37: Streamlines for a=0.5, M=, b=0.5, ϕ=0; d=1; $\overline{Q}$=1.8, λ_1=0.1, β=0.0, a) Sinusoidal wave, b) Triangular wave and c) Trapezoidal wave.

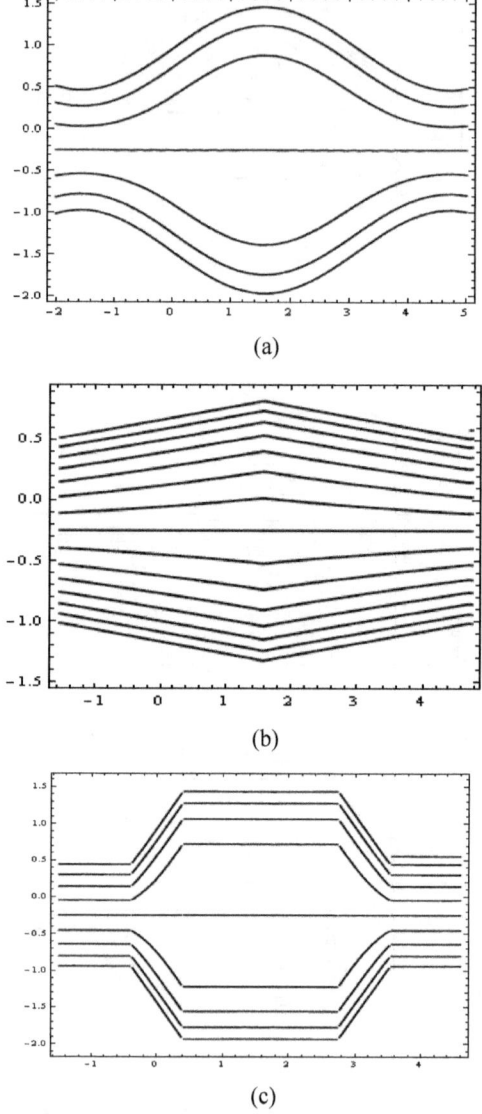

(a)

(b)

(c)

Figure 1.38: Streamlines for a=0.5, M=1, b=0.5; ϕ=0, d=1, $\overline{Q}$=1; λ_1=0, β=0.4, a)
Sinusoidal wave, b) Triangular wave and c) Trapezoidal wave.

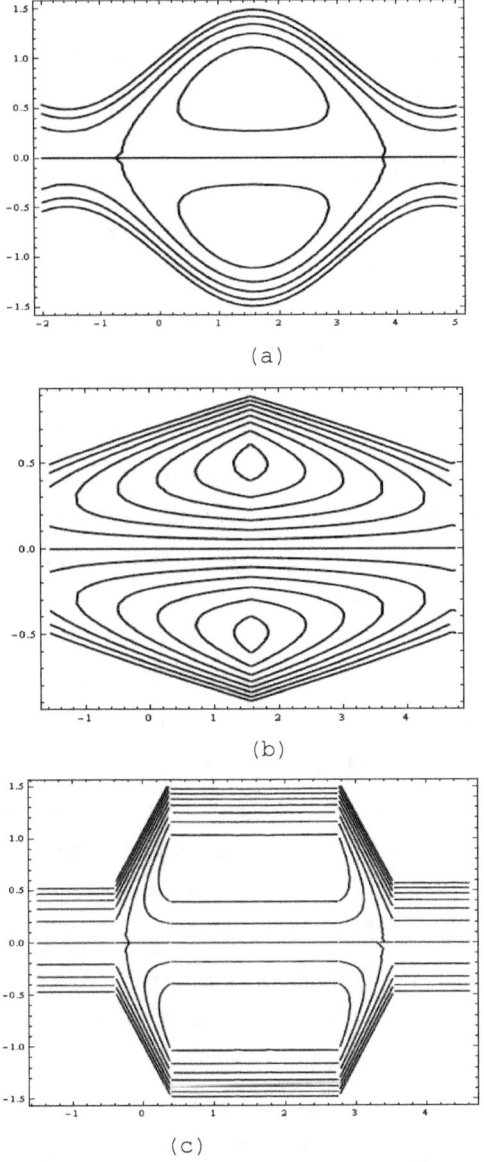

Figure 1 39: Streamlines for a=0.5, M=1, b=0.5, Φ=0; d=1, $\overline{Q}$=1, λ=1, β=0.4, a) Sinusoidal wave, b) Triangular wave and c) Trapezoidal

2.1 Introduction

Peristaltic transport of a non-Newtonian biofluid through a channel bounded by permeable walls is of considerable importance in biology and medicine. Peristalsis is an inherent property of many of the smooth muscle tubes such as the gastrointestinal tract, bile duct. Even though peristalsis existed very well in physiology, its relevance came about mainly through the works of Kill (1957) and Boyarsky (1964). Later several mathematical and experimental models have been developed to understand the fluid mechanical aspect of peristaltic motion. The mathematical model obtained by a train of periodic sinusoidal waves in an infinitely long two-dimensional symmetric channel and tube containing a Newtonian fluid have been investigated by Shapiro *et al.* (1969). Later Shukla and Guptha(1982), Srivastva and Srivastava (1984) and many others solved various problems in peristaltic pumping of Newtonian/non-Newtonian fluids. Many of these models explain the basic fluid mechanics aspects of peristalsis, namely the characteristics of pumping, trapping and reflux. These models are developed in two ways, one by restricting to small peristaltic wave amplitude with arbitrary Reynolds number and the other by lubrication theory in which the fluid inertia and small curvature are neglected without any restrictions of wave amplitude. The problems are investigated either in a fixed frame of reference or in a wave frame of reference moving with constant velocity of the wave simplifying the study to a case with stationary wavy walls. The accuracy of these models has been investigated numerically and experimentally by Takabataka and Ayukawa (1982) and several others. Brasseur *et al.* (1987) discussed the influence of a peripheral layer of different viscosity on peristaltic pumping with Newtonian fluids. Vajravelu *et al.* (1987) observed Bingham fluid flow through a circular pipe with permeable wall. Hayat *et al.* (2007) investigated peristaltic flow of a micropolar fluid in a channel with different wave forms. It is clear that most of the physiological fluids (for example, blood) can not be described by Newtonian model. So these are discussed using non-Newtonian models. Several non-Newtonian models are proposed by various researchers. Among them Williamson model may be expected to explain the features of a physiological fluid as this model is nonlinear and Newtonian fluid model can be deduced as a special case from this model. Irene and Giambattista (2007) discussed perturbation solutions for pulsatile flow of a non-Newtonian Williamson fluid in a rock fracture. Prasanna Hariharan *et al.* (2008) examined the peristaltic transport of non-Newtonian fluid in a diverging tube with different wave forms. Nadeem and Akram (2010) discussed peristaltic flow of a Williamson fluid in an asymmetric channel.

Most of the biological organs contain a permeable layer attached to the wall. For example, blood vessel consists of tissue layer which is modeled as a porous layer by many researchers. In view of these facts, the influence of permeability on peristaltic transport of Williamson fluid is investigated. The velocity and the stream function are determined. The pressure rise per one wave length is calculated and the effects of various physical parameters on the pumping characteristics are analyzed.

2.2 Williamson fluid model

For an incompressible fluid the balance of mass and momentum are given by

$$div V = 0 ,$$ (2.1)

$$\rho \frac{dV}{dt} = div S + \rho f ,$$ (2.2)

Where ρ is the density, V is the velocity vector, S is the Cauchy stress tensor and f represents the specific body force and $\frac{d}{dt}$ represents the material time derivative. The constitutive equation for Williamson fluid is given by

$$S = -PI + \tau ,$$ (2.3)

$$\tau = -\left[\mu_\infty + (\mu_0 + \mu_\infty)(1 \Gamma \bar{y})^{-1} \right] \bar{y}.$$ (2.4)

in which $-PI$ is the spherical part of the stress due to constraint of incompressibility, τ is the extra stress tensor, μ_∞ is the infinite shear rate viscosity, μ_0 is the zero shear rate viscosity, Γ is the time constant and $\dot{y}$ is defined as

$$\bar{y} = \sqrt{\frac{1}{2} \sum_i \sum_j \bar{y}_{ij} \bar{y}_{ji}} = \sqrt{\frac{1}{2} \Pi} .$$ (2.5)

Here Π is the second invariant strain tensor. We consider the constitutive equation (2.4), the case for which $\mu_\infty = 0$ $and\,\Gamma\dot{y}<1$.The component of extra stress tensor therefore, can be written as

$$\bar{\tau} = -\mu_0\left[(1-\Gamma\bar{\dot{y}})^{-1}\right]\bar{\dot{y}} = -\mu_0\left[(1+\Gamma\bar{\dot{y}})\right]\bar{\dot{y}}. \tag{2.6}$$

2.3 Mathematical formulation

Let us consider the peristaltic transport of an incompressible Williamson fluid in a two dimensional channel of width $\bar{d}_1+\bar{d}_2$ (Figure 2.1). The flow is generated by sinusoidal wave trains propagating with constant speed c along the permeable walls of the channel. The geometry of the wall surfaces is defined as

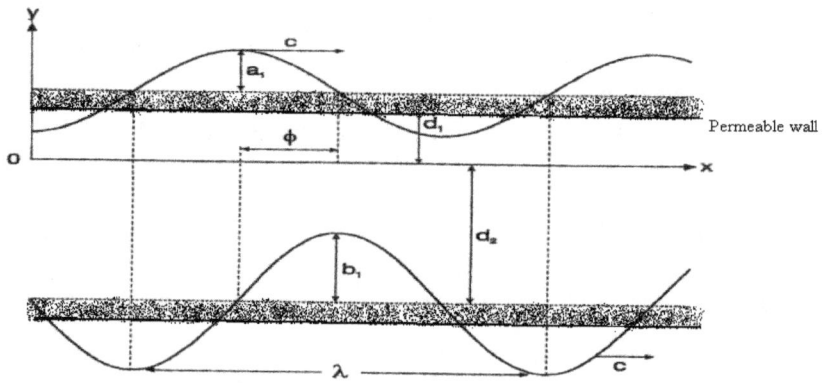

Figure 2.1.Schematic diagram of a two-dimensional asymmetric channel

$$Y = H_1 = \bar{d}_1 + \bar{a}_1\cos\left[\frac{2\pi}{\lambda}(\bar{X}-c\bar{t})\right]\quad\ldots\ldots\text{ upper wall,} \tag{2.7}$$

$$Y = H_1 = \bar{d}_1 + \bar{b}_1\cos\left[\frac{2\pi}{\lambda}(\bar{X}-c\bar{t})\right]\quad\ldots\ldots\text{lower wall.} \tag{2.8}$$

where $\bar{a}_1$ and $\bar{b}_1$ are the amplitudes of the waves, λ is the wave length, $\bar{d}_1+\bar{d}_2$ is the width of the channel, c is the velocity of propagation, $\bar{t}$ is the time and $\bar{X}$ is the direction of wave propagation. The phase difference ϕ varies in the range $0\le\phi\le\pi$ in which $\phi = 0$ corresponds to symmetric channel with waves out of phase and $\phi = \pi$ the waves are in phase, further $\bar{a}_1$, $\bar{b}_1$, $\bar{d}_1$, $\bar{d}_2$ and ϕ satisfies the condition.

$$\overline{a}_1^2 + \overline{b}_1^2 + 2\overline{a}_1 b_1 \cos\phi \le (\overline{d}_1 + \overline{d}_2)^2.$$

The equations governing the flow of a Williamson fluid are given by

$$\frac{\partial \overline{U}}{\partial \overline{X}} + \frac{\partial \overline{V}}{\partial \overline{Y}} = 0, \tag{2.9}$$

$$\rho\left(\frac{\partial \overline{U}}{\partial \overline{t}} + \overline{U}\frac{\partial \overline{U}}{\partial \overline{X}} + \overline{V}\frac{\partial \overline{U}}{\partial \overline{Y}}\right) = -\frac{\partial \overline{p}}{\partial \overline{X}} + \frac{\partial \overline{\tau}_{\overline{X}\overline{X}}}{\partial \overline{Y}} + \frac{\partial \overline{\tau}_{\overline{X}\overline{Y}}}{\partial \overline{Y}}, \tag{2.10a}$$

$$P\left(\frac{\partial \overline{V}}{\partial \overline{t}} + \overline{U}\frac{\partial \overline{U}}{\partial \overline{X}} + \overline{V}\frac{\partial \overline{V}}{\partial \overline{Y}}\right) = -\frac{\partial \overline{p}}{\partial \overline{Y}} + \frac{\partial \overline{\tau}_{\overline{X}\overline{X}}}{\partial \overline{Y}} + \frac{\partial \overline{\tau}_{\overline{Y}\overline{Y}}}{\partial \overline{Y}}. \tag{2.10b}$$

The flow in the permeable wall is described by Darcy's law. The velocity in the permeable wall is given by

$$\overline{u}_D = -\frac{k}{\mu_0}\frac{\partial \overline{p}}{\partial \overline{x}}. \tag{2.11}$$

where k is the permeability of the wall.

We introduce a wave frame $(\overline{x}, \overline{y})$ moving with velocity c away from the fixed frame $(\overline{X}, \overline{Y})$ by the transformation

$$\overline{x} = \overline{X} - c\overline{t}, \overline{y} = \overline{Y} \quad \overline{u} = \overline{U} - c, \quad \overline{v} = \overline{V} \quad and \quad \overline{P}(x) = \overline{P}(X, t). \tag{2.12}$$

We define the following non-dimensional quantities

$$x = \frac{\overline{x}}{\lambda}, \quad y = \frac{\overline{y}}{d_1}, \quad u = \frac{\overline{u}}{c}, \quad v = \frac{\overline{v}}{c}, \quad h_1 = \frac{\overline{h}_1}{d_1}, \quad u_D = \frac{\overline{u}_D}{c}, \quad Da = \frac{k}{\overline{d}_1},$$

$$h_2 = \frac{\overline{h}_2}{\overline{d}_2}, \quad \tau_{xx} = \frac{\lambda}{\mu_0 c}\tau_{\overline{x}\overline{y}}, \quad \tau_{xy} = \frac{\overline{d}_1}{\mu_0 c}\tau_{\overline{x}\overline{y}}, \quad \tau_{yy} = \frac{\overline{d}_1}{\mu_0 c}\tau_{\overline{x}\overline{y}},$$

$$\delta = \frac{\overline{d}_1}{\lambda}, \quad \mathrm{Re} = \frac{\rho c \overline{d}_1}{\mu_0}, \quad We = \frac{\Gamma c}{d_1}, \quad P = \frac{\overline{d}_1^2}{c\lambda\mu_0}\overline{P}, \quad \dot{\gamma} = \frac{\overline{\gamma}\overline{d}_1}{c}.$$

Using the above non – dimensional quantities in equations (2.10a) and (2.10b) the resulting equations in terms of stream function $\psi\left(u = \frac{\partial\psi}{\partial y}, v = -\delta\frac{\partial\psi}{\partial x}\right)$ can be written as

$$\delta \operatorname{Re}\left[\left(\frac{\partial \Psi}{\partial y}\frac{\partial}{\partial x}-\frac{\partial \Psi}{\partial x}\frac{\partial}{\partial y}\right)\frac{\partial \Psi}{\partial y}\right]=-\frac{\partial P}{\partial x}+\delta^2\frac{\partial \tau_{xx}}{\partial x}+\frac{\partial \tau_{xy}}{\partial y},$$

(2.13a)

$$-\delta^3 \operatorname{Re}\left[\left(\frac{\partial \Psi}{\partial y}\frac{\partial}{\partial x}-\frac{\partial \Psi}{\partial x}\frac{\partial}{\partial y}\right)\frac{\partial \Psi}{\partial x}\right]=-\frac{\partial P}{\partial y}+\delta^2\frac{\partial \tau_{xy}}{\partial x}+\delta\frac{\partial \tau_{yy}}{\partial y}.$$ (2.13b)

$$u_D=-Da\frac{\partial p}{\partial x}$$ (2.14)

where

$$\tau_{xx}=2\left[1+We\dot\gamma\right]\frac{\partial^2\Psi}{\partial x\partial y},$$

$$\tau_{xy}=\left[1+We\dot\gamma\right]\left(\frac{\partial^2\Psi}{\partial y^2}-\delta^2\frac{\partial^2\Psi}{\partial x^2}\right),$$

$$\tau_{yy}=-2\delta\left[1+We\dot\gamma\right]\frac{\partial^2\Psi}{\partial x\partial y},$$

$$\dot\gamma=\left[2\delta^2\left(\frac{\partial^2\Psi}{\partial x\partial y}\right)^2+\left(\frac{\partial^2\Psi}{\partial y^2}-\delta^2\frac{\partial^2\Psi}{\partial x^2}\right)^2+2\delta^2\left(\frac{\partial^2\Psi}{\partial x\partial y}\right)^2\right]^{\frac{1}{2}}.$$

in which δ, Re and We represent the wave, Reynolds and Weissenberg numbers, respectively. Under the assumptions of long wavelength $\delta \ll 1$ and low Reynolds number, neglecting the terms of order δ and higher, equations (2.13) and (2.14) take the form

$$\frac{\partial P}{\partial x}=\frac{\partial}{\partial y}\left[\left(1+We\frac{\partial^2\Psi}{\partial y^2}\right)\frac{\partial^2\Psi}{\partial x\partial y}\right],$$ (2.15)

$$\frac{\partial P}{\partial y}=0.$$ (2.16)

Eliminating of pressure from equations (15) and (16), yield

$$\frac{\partial^2}{\partial y^2}\left[\left(1+We\frac{\partial^2\Psi}{\partial y^2}\right)\frac{\partial^2\Psi}{\partial y^2}\right]=0.$$ (2.17)

The dimensionless mean flow $\bar{Q}$ is defined by

$$\bar{Q} = q + 1 + d \,, \tag{2.18}$$

in which

$$F = \int_{h_2(x)}^{h_1(x)} \frac{\partial \Psi}{\partial y}\, dy = \Psi\big(h_1(x) - h_2(x)\big), \tag{2.19}$$

where

$$h_1(x) = 1 + a\cos 2\pi x, \quad h_2(x) = -d - b\cos(2\pi x + \phi). \tag{2.20}$$

The boundary conditions in terms of stream function Ψ are defined as

$$\psi = \frac{q}{2}\, at\ y = h_1(x), \tag{2.21}$$

$$\psi = \frac{-q}{2}\, at\ y = h_2(x). \tag{2.22}$$

$$\frac{\partial \psi}{\partial y} + \beta \frac{\partial^2 \psi}{\partial y^2} = -1\, at\ y = h_1(x), \tag{2.23}$$

$$\frac{\partial \psi}{\partial y} - \beta \frac{\partial^2 \psi}{\partial y^2} = -1\, at\ y = h_2(x). \tag{2.24}$$

where q is the flux, $\beta = \dfrac{\sqrt{k}}{\alpha d_1} = \dfrac{\sqrt{Da}}{\alpha}$ is permeability parameter including slip, Da is

the Darcy number, α is the slip parameter and k is the permeability parameter α.

The boundary conditions (2.23) and (2.24) are introduced at the permeable walls of the channel following Saffman (1971). When the flow takes place past a permeable bed, it is well known that the usual no-slip condition is not valid at the nominal surface of the porous bed (Beavers and Joseph, 1967). The slip at the permeable wall is presented through a slip condition formulated by Saffman which is an improved condition to the Beavers and Joseph slip condition.

2.4 Perturbation solution

Since, equation (2.17) is non-linear equation, its exact solution is not possible, therefore, we employ the regular perturbation technique to find the solution.

For perturbation solution, we expand Ψ, q and P as

$$\Psi = \Psi_0 + We\Psi_1 + O\left(We^2\right), \qquad (2.25)$$

$$q = q_0 + Weq_1 + O\left(We^2\right), \qquad (2.26)$$

$$P = P_0 + WeP_1 + O\left(We^2\right). \qquad (2.27)$$

Substituting the above expressions in equations (2.17) and (2.18) and boundary conditions (2.21) to (2.24), we get the following system

2.4.1 System of order We^0

$$\frac{\partial^4 \Psi_0}{\partial y^4} = 0, \qquad (2.28)$$

$$\frac{\partial P_0}{\partial x} = \frac{\partial^3 \Psi_0}{\partial y^3}, \qquad (2.29)$$

$$\psi_0 = \frac{q_0}{2} \text{ at } y = h_1(x), \qquad (2.30)$$

$$\psi_0 = \frac{-q_0}{2} \text{ at } y = h_2(x), \qquad (2.31)$$

$$\frac{\partial \psi_0}{\partial y} + \beta \frac{\partial^2 \psi_0}{\partial y^2} = -1 \text{ at } y = h_1(x), \qquad (2.32)$$

$$\frac{\partial \psi_0}{\partial y} - \beta \frac{\partial^2 \psi_0}{\partial y^2} = -1 \text{ at } y = h_2(x). \qquad (2.33)$$

2.4.2 System of order We^1

$$\frac{\partial^4 \Psi_1}{\partial y^4} = -\frac{\partial^2}{\partial y^2}\left(\frac{\partial^2 \Psi_0}{\partial y^2}\right)^2, \qquad (2.34)$$

$$\frac{\partial P_1}{\partial x} = \frac{\partial^3 \Psi_1}{\partial y^3} + \frac{\partial}{\partial y}\left(\frac{\partial^2 \Psi_0}{\partial y^2}\right)^2, \qquad (2.35)$$

$$\psi_1 = \frac{q_1}{2} \text{ at } y = h_1(x), \qquad (2.36)$$

$$\psi_1 = \frac{-q_1}{2} \text{ at } y = h_2(x), \qquad (2.37)$$

$$\frac{\partial \psi_1}{\partial y} + \beta \frac{\partial^2 \psi_1}{\partial y^2} = -1 \quad at \ y = h_1(x),$$ (2.38)

$$\frac{\partial \psi_1}{\partial y} - \beta \frac{\partial^2 \psi_1}{\partial y^2} = -1 \quad at \ y = h_2(x).$$ (2.39)

2.4.3 Solution for system of order We^0

Solution of equation (2.28) satisfying the boundary conditions (2.30) to (2.33) can be written as

$$\psi_0 = C_1 \frac{y^3}{3!} + C_2 \frac{y^2}{2!} + C_3 y + C_4$$ (2.40)

where

$$C_1 = \frac{-12\left[q_0 + h_1 - h_2\right]}{(h_1 - h_2)^2 \left[h_1 - h_2 + 6\beta\right]},$$

$$C_2 = \frac{6(q_0 + h_1 - h_2)(h_1 + h_2)}{(h_1 - h_2)^2 \left[h_1 - h_2 + 6\beta\right]},$$

$$C_3 = \frac{-(h_1 - h_2)\left[(h_1 - h_2)^2 - 6(q_0\beta - h_1 h_2)\right] - 6q_0 h_1 h_2}{(h_1 - h_2)^2 (h_1 - h_2 + 6\beta)},$$

$$C_4 = \frac{\frac{q_0}{2} + h_1(q_0 + h_1 - h_2)\left[2 - 3h_1(h_1 + h_2)\right] + \left[(h_1 - h_2)^2 - 6(q_0\beta - h_1 h_2)\right] + 6q_0 h_1 h_2}{(h_1 - h_2)^2 (h_1 - h_2 + 6\beta)}.$$

The axial pressure gradient at this order is

$$\frac{\partial p_0}{\partial x} = \frac{-12\left[q_0 + h_1 - h_2\right]}{(h_1 - h_2)^2 \left[h_1 - h_2 + 6\beta\right]}.$$ (2.41)

For one wavelength the integration of equation (2.41), yield

$$\Delta P_{\lambda 0} = \int_0^1 \frac{dp_0}{dx} \, dx.$$ (2.42)

2.4.4 Solution for system of order We^1

Substituting the zeroth-order solution (2.40) into (2.34), the solution of the resulting problem satisfying the boundary conditions take following form:

$$\psi_1 = F_1 \frac{y^3}{3!} + F_2 \frac{y^2}{2!} + F_3 y + F_4 - Ay^4.$$ (2.43)

61

where

$$F_1 = \frac{-12\left[(q_1 + h_1 - h_2)\right] - A_{12}(h_1 - h_2)}{(h_1 - h_2)^2 \left[h_1 - h_2 + 6\beta\right]},$$

$$F_2 = \frac{6(h_1 + h_2)\,(q_1 + h_1 - h_2) + A_{13}}{(h_1 - h_2)^2 \left[h_1 - h_2 + 6\beta\right]},$$

$$F_3 = \frac{6(q_1 + h_1 - h_2)\left[(\beta(h_1 - h_2) - h_1 h_2] + A_{14}\right.}{(h_1 - h_2)^2 \left[h_1 - h_2 + 6\beta\right]},$$

$$F_4 = \frac{q_1}{2}\frac{(q_1 + h_1 - h_2)\left[(-h_1^3 + 3h_1^2 h_2 - 6\beta h_1(h_1 - h_2)\right] + A_{15}}{(h_1 - h_2)^2 \left(h_1 - h_2 + 6\beta\right)},$$

$$A = \frac{-288\left(q_0 + h_1 - h_2\right)^2}{4!(h_1 - h_2)^4 \,(h_1 - h_2 + 6\beta)^2}.$$

The axial pressure gradient at this order is

$$\frac{\partial p_1}{\partial x} = \frac{\begin{bmatrix} -12(q_1 + h_1 - h_2)\left[(h_1 - h_2)(h_1 - h_2 + 6\beta) + 12(q_0 + h_1 - h_2)(h_1 + h_2)\right] \\ -A_{12}(h_1 - h_2)(h_1 - h_2 + 6\beta) \end{bmatrix}}{(h_1 - h_2)^2 \,(h_1 - h_2 + 6\beta)^2}. \quad (2.44)$$

Integrating above equation over one wavelength, we get

$$\Delta P_{\lambda 1} = \int_0^1 \frac{dp_1}{dx}\,dx. \quad (2.45)$$

Summarizing the perturbation result for small parameter *We*, the expression for stream function, the velocity and pressure gradient can be written as

$$\Psi = C_1 \frac{y^3}{3!} + C_2 \frac{y^2}{2!} + C_3 y + C_4 + We\left(F_1 \frac{y^3}{3!} + F_2 \frac{y^2}{2!} + F_3 y + F_4 - Ay^4 \right), \quad (2.46)$$

$$u = C_1 \frac{y^2}{2!} + C_2 y + C_3 + We\left(F_1 \frac{y^2}{2!} + F_2 y + F_3 - 4Ay^3 \right), (2.47)$$

$$\frac{dp}{dx} = \frac{-12\left[q_0 + h_1 - h_2\right]}{(h_1 - h_2)^2 \left[h_1 - h_2 + 6\beta\right]} + We\left[\frac{\begin{bmatrix} -12(q_1 + h_1 - h_2)\left[(h_1 - h_2)(h_1 - h_2 + 6\beta) + 12(q_0 + h_1 - h_2)(h_1 + h_2)\right] \\ -A_{12}(h_1 - h_2)(h_1 - h_2 + 6\beta) \end{bmatrix}}{(h_1 - h_2)^2 \,(h_1 - h_2 + 6\beta)^2} \right] \quad (2.48)$$

The non-dimensional pressure rise over one wavelength ΔP_λ for the axial velocity is

$$\Delta P_\lambda = \int_0^1 \frac{dp}{dx} dx .$$
(2.49)

where $\frac{dp}{dx}$ is defined in equation (2.48)

The frictional force, at $y = h_1$ and $y = h_2$ denoted by

$$F_{\lambda 1} = \int_0^1 -h_1^2 \left(\frac{dp}{dx} \right) dx ,$$
(2.50)

$$F_{\lambda 2} = \int_0^1 -h_2^2 \left(\frac{dp}{dx} \right) dx .$$
(2.51)

2.5 Expressions for wave shape:

The non-dimensional expressions for the five considered wave forms are given by the following equations:

1. Sinusoidal wave:

$$h(x) = 1 + a \sin(x),$$
(2.52)

2. Triangular wave:

$$h(x) = 1 + a \left\{ \frac{8}{\pi^3} \sum_{m=1}^{\infty} \frac{(-1)^{m+1}}{(2m-1)^2} \sin[(2m-1)x] \right\},$$
(2.53)

3. Square wave:

$$h(x) = 1 + a \left\{ \frac{4}{\pi} \sum_{m=1}^{\infty} \frac{(-1)^{m+1}}{(2m-1)} \cos[(2m-1)x] \right\},$$
(2.54)

4. Saw tooth wave:

$$h(x) = 1 + a \left\{ \frac{8}{\pi^3} \sum_{m=1}^{\infty} \frac{\sin[2m\pi x]}{m} \right\},$$
(2.55)

5. Trapezoidal wave:

$$h(x) = 1 + a \left\{ \frac{32}{\pi^2} \sum_{m=1}^{\infty} \frac{\sin \frac{\pi}{8} (2m-1)}{(2m-1)^2} \sin[(2m-1)x] \right\}.$$
(2.56)

Appendix

$$A = \frac{-288\left(q_0 + h_1 - h_2\right)^2}{4!(h_1 - h_2)^4\,(h_1 - h_2 + 6\beta)^2},$$

$$A_{11} = \frac{A\left[4(h_1^3 - h_2^3) + 6\beta(h_1^2 + h_2^2)\right]}{(h_1 - h_2 + 2\beta)},$$

$$A_{12} = A\left[(h_1^2 + h_2^2)(h_1 + h_2) - 4h_1^3 - 6\beta h_1^2\right] - \frac{A_{11}}{2}(h_2 - h_2 - 2\beta),$$

$$A_{13} = -6A_{12}(h_1 - h_2)(h_1 + h_2) - A_{11}(h_1 - h_2)^2(h_1 - h_2 + 6\beta),$$

$$A_{14} = \frac{-6A_{12}(h_1 - h_2)(h_1 + 2\beta h_1) - A_{13}(h_1 + \beta) - (1 + 6\beta Ah_1^2)(h_1 - h_2)^2(h_1 - h_2 + 6\beta)}{(h_1 - h_2)^2(h_1 - h_2 + 6\beta)},$$

$$A_{15} = -2A_{12}(h_1 - h_2)h_1^3 - A_{14}h_1 - Ah_1^4(h_1 - h_2)^2(h_1 - h_2 + 6\beta).$$

2.6 Results and discussion:

The variation of velocity u with y is calculated from equation (2.47) different values of the average volume flow rate $\overline{Q}$ with x=1, a=0.6, b=0.8, d=1.2, $\phi = \frac{\pi}{3}$, β =0.5 and We =0.01 and is depicted in figure (2.2). It is observed that velocity u increases by increasing the average flow rate $\overline{Q}$.

The relationship between velocity u with y is shown for different values of the permeability parameter including slip β with x=1, a=0.6, b=0.8, d=1.2, $\phi = \frac{\pi}{3}$, $\overline{Q}$ =2 and We =0.01 in figure (2.3). It is concluded that the velocity u increases with increasing permeability parameter including slip β act as an increasing resistant against the flow in the in the central part of the channel.

In figure (2.4) the relation between velocity u with y is plotted with different values of Weissenberg number We with x=1, a=0.6, b=0.8, d=1.2, $\phi = \frac{\pi}{3}$, $\overline{Q}$=2 and β =0.5. It is observed that increase in the Weissenberg number We decrease velocity.

The variation between velocity a u with y is drawn in figure (2.5) for different values of the phase difference ϕ with x=1, a=0.6, b=0.8, d=1.2, $\overline{Q}$=2, β =0.5 and We =0.01. It is concluded that the velocity decrease by increasing phase difference ϕ.

The variation on pressure rise ΔP_λ with the average volume flow rate $\overline{Q}$ is calculated from equation (2.49) and is drawn figure (2.6) for different values of permeability parameter including slip β with a=0.5, b=0.5, d=1.2, $\phi = \dfrac{\pi}{3}$ and We =0.01. We conclude that for values of $\overline{Q}$ between 0.5 and 0.6, the pumping curves intersect at a point (0.55, -0.2). For a given mean flux $\overline{Q}$, the pressure rise ΔP_λ increases with increasing permeability parameter including slip β below this point and opposite behaviour is observed above this point.

The relation between pressure rise ΔP_λ with the average volume flow rate $\overline{Q}$ is plotted figure (2.7) for different values of phase difference ϕ with a=0.5, b=0.5, d=1, β =0.01 and We =0.01. We observe that for values of $\overline{Q}$ between 1.2 and 1.4, the pumping curves intersect at a point (1.21, -1.8). For a given average volume flow rate $\overline{Q}$, the pressure rise ΔP_λ increases with decreasing phase difference ϕ above this point and opposite behaviour is observed below this point.

Figure (2.8) shows the relation between pressure rise ΔP_λ with the average volume flow rate $\overline{Q}$ for various values of Weissenberg number We with a=0.5, b=0.5, d=1.1, $\phi = \dfrac{\pi}{3}$ and β =0.01. It is observed pressure rise ΔP_λ decreases with increasing the Weissenberg number We .

The relationship between pressure rise ΔP_λ with the average volume flow rate $\overline{Q}$ is drawn figure (2.9) for different values of channel width d with a=0.5, b=0.5, $\phi = \dfrac{\pi}{3}$, β =0.01 and We =0.01. We conclude that for values of $\overline{Q}$ between 0.5 and 1, the pumping curves intersect at a point (0.55, 0). For a given average volume flow rate $\overline{Q}$, the pressure rise ΔP_λ increases with increasing channel width d below this point and opposite behaviour is observed above this point.

The variation of pressure rise ΔP_λ with the average volume flow rate $\overline{Q}$ is depicted figure (2.10) for different values of wave amplitude a with b=0.5, d=1, $\phi = \frac{\pi}{3}$, β =0.01 and We =0.01. We infer that for values of $\overline{Q}$ between 0.5 and 1, the pumping curves intersect at a point (0.9, -3.8). For a given average volume flow rate $\overline{Q}$, the pressure rise ΔP_λ decreases with increasing wave amplitude a below this point and opposite behaviour is observed above this point.

The relation between pressure rise ΔP_λ with the average volume flow rate $\overline{Q}$ is depicted figure (2.11) for different values of wave amplitude b with a=0.5, d=1, $\phi = \frac{\pi}{3}$, β =0.01 and We =0.01. We found that for values of $\overline{Q}$ between 0.5 and 1, the pumping curves intersect at a point (0.9, -3). For a given average volume flow rate $\overline{Q}$, the pressure rise ΔP_λ decreases with increasing wave amplitude b below this point and opposite behaviour is concluded above this point.

The variation of frictional forces $F_{\lambda 1}$ and $F_{\lambda 2}$ with $\overline{Q}$ is shown figures (2.12) to (2.17). We observe that the frictional force shows opposite behaviour to that of pressure rise for the corresponding variations in the physical parameters ϕ, β and We.

The variation of pressure gradient $\frac{dp}{dx}$ with x is calculated from equation (2.48) for different values of phase difference ϕ with a=0.5, b=0.5, d=1, $\phi = \frac{\pi}{3}$, $\overline{Q}$=-1, β =0.01 and We =0.01 and is plotted in figure (2.18). It is observed that pressure gradient $\frac{dp}{dx}$ decreases with the increase in the phase difference ϕ and point of maximum amplitudes decreases with increasing ϕ.

In figure (2.19) the relation between pressure gradient $\frac{dp}{dx}$ and x is plotted at different values of the mean flux $\overline{Q}$ with a=0.5, b=0.5, d=1, $\phi = \frac{\pi}{3}$, β=0.01 and We =0.01. It is noticed that the pressure gradient $\frac{dp}{dx}$ increases with increasing mean flux $\overline{Q}$.

The variation between pressure gradient $\frac{dp}{dx}$ with x is shown in figure (2.20) for different values of permeability parameter including slip β with a=0.5, b=0.5, d=1, $\overline{Q}$=-1, $\phi = \frac{\pi}{3}$ and We =0.01. It is concluded that the pressure gradient $\frac{dp}{dx}$ increases with decreasing permeability parameter including slip β.

The relation between pressure gradient $\frac{dp}{dx}$ and x is drawn in figure (2.21) for different values of Weissenberg number We with a=0.5, b=0.5, d=1, $\overline{Q}$=-1, $\phi = \frac{\pi}{3}$ and β=0.01.It is found that the pressure gradient $\frac{dp}{dx}$ decreases with increasing Weissenberg number We .

From figure (2.22) the variation pressure gradient $\frac{dp}{dx}$ with x is plotted for different channel width d with a=0.5, b=0.5, $\overline{Q}$=-2, $\phi = \frac{\pi}{3}$, β=0.01and We =0.01. It is concluded that the pressure rise deceases by increasing channel width d.

2.6.1 Trapping phenomena

Another interesting phenomenon in peristaltic motion is trapping. It is basically the formation of an internally circulating bolus of fluid by closed stream lines. The trapped bolus will be pushed ahead along the peristaltic waves.

The stream lines are plotted in figure (2.23) for different amplitude of the wave a with b=0.5, d=1, β=0.01, ϕ=0, $\overline{Q}$=2 and We =0.01. It is observed that when wave amplitude a is equals to zero, peristaltic waves exists only at the lower wall and for a>0 the size of the trapping bolus increases in the upper half of the channel and decreases in the lower half of the channel with the increase in amplitude of the wave a.

The stream lines are plotted in figure (2.24) for different amplitude of the wave b with a=0.5, d=1, β=0.01, φ=0, $\overline{Q}$=2 and We =0.01. It is observed that when wave amplitude b is equals to zero, peristaltic waves exists only at the upper wall and for b>0 the size of the trapping bolus decreases in the upper half of the channel and increases in the lower half of the channel with the increasing in amplitude of the wave b.

The stream lines are plotted in figure (2.25) for various values of the phase difference φ with a=0.5, b=0.5, d=1, β=0.01, $\overline{Q}$=2 and We =0.01. It is concluded that the volume of the trapping bolus appearing in the central region for φ=0, moves towards left and decreases in volume as φ increases.

The stream lines are drawn in figure (2.26) with different values of mean flux $\overline{Q}$ at a=0.5, b=0.5, d=1, β=0.01, φ=0 and We =0.01. It is noticed that the size of the trapping bolus increases in the lower half of the channel by increasing mean flux $\overline{Q}$ and bolus disappears for $\overline{Q}$=1.

The stream lines are depicted in the wave frame for pumping for different values of channel width d in figure (2.27) with a=0.5, b=0.5, β=0.01, φ=0, $\overline{Q}$=2 and We =0.01. It is observed that the size of the trapping bolus decreases with increasing channel width d.

The stream lines are plotted in the wave frame for pumping for different values of Weissenberg number We in figure (2.28) with a=0.5, b=0.5, β=0.01, φ=0, $\overline{Q}$=2 and d=1. It is observed that the size of the trapping bolus decreases in the upper half of the channel and increases in the lower half of the channel with increasing Weissenberg number We .

The stream lines are drawn for various values of the permeability parameter including slip β and are shown in figure (2.29) with a=0.5, b=0.5, d=1, φ=0, $\overline{Q}$=2 and We =0.01. It is observed that the volume the trapping bolus decreases by increasing permeability parameter including slip β and the trapping bolus disappears for β=0.6.

Stream lines are plotted in figures (2.30) and (2.31) for different values of β with a=0.5, b=0.5, d=1, ϕ=0, $\bar{Q}$=2 and We =0.01 for following wave forms a)sinusoidal wave, b) triangular wave, c) square wave, d) sawtooth wave and e) Trapezoidal wave. We observed that the size of the bolus decreases with increasing β for the wave forms considered.

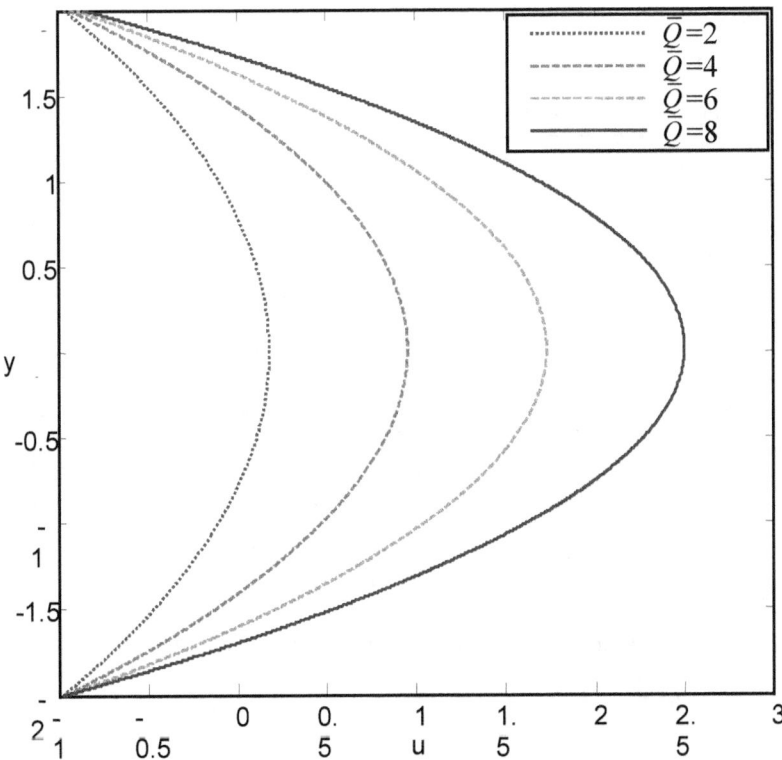

Figure 2.2: The velocity profiles with a=0.6, b=0.8, d=1.2, x=1, $\phi = \frac{\pi}{3}$, β=0.5and We =0.01.

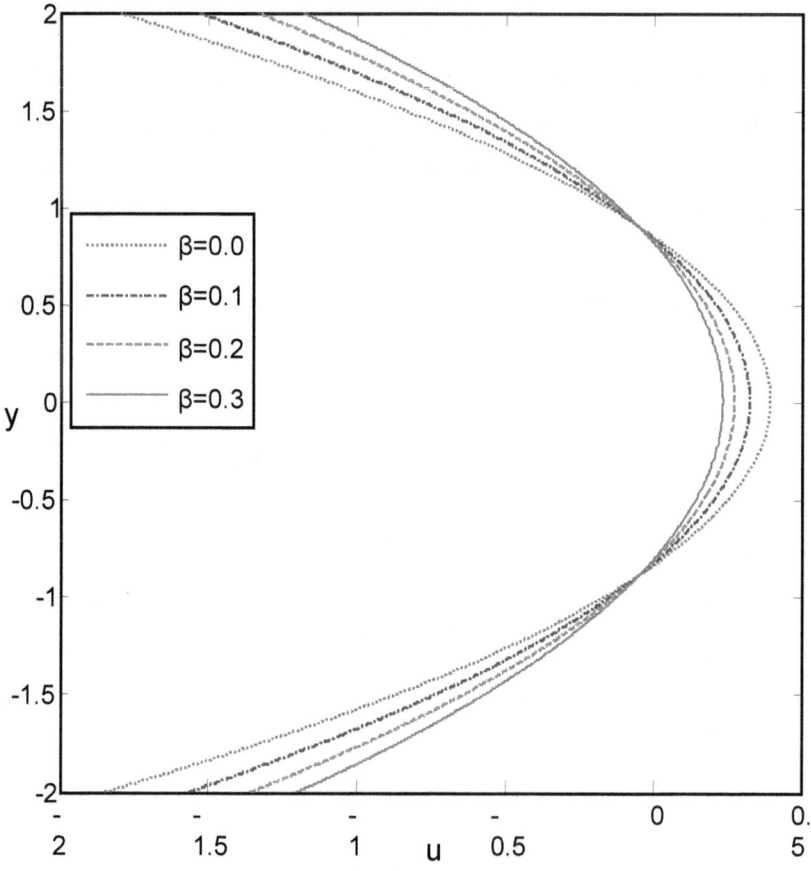

Figure 2.3: The velocity profiles with a=0.6, b=0.8, d=1.2, x=1, $\overline{Q}$=2, $\phi = \dfrac{\pi}{3}$ and We =0.01.

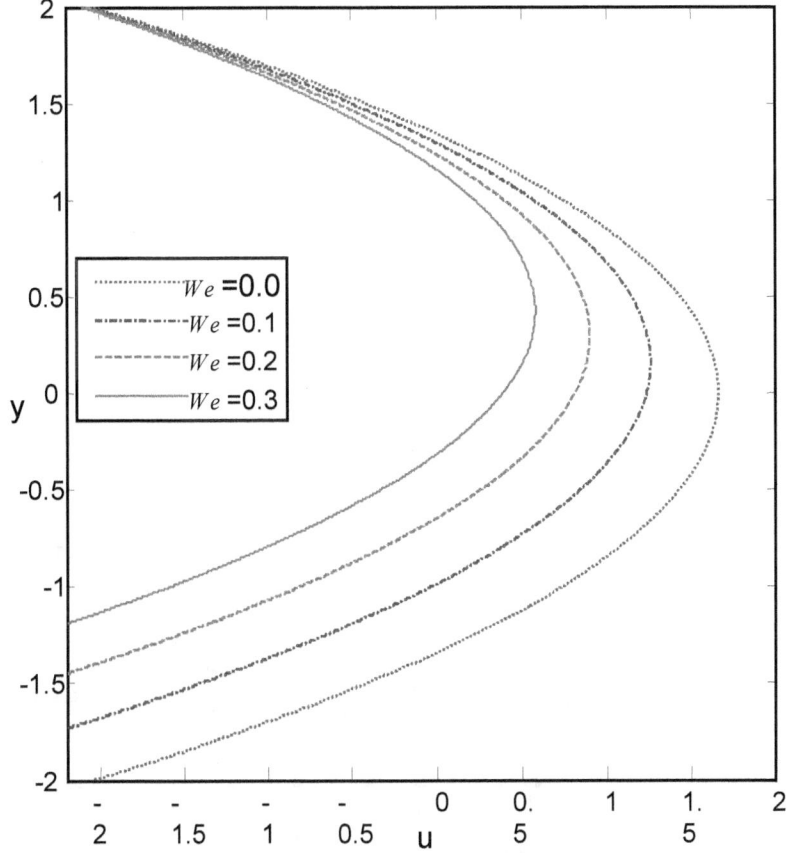

Figure 2.4: The velocity profiles with a=0.6, b=0.8, d=1.2, x=1, $\overline{Q}$ =2, $\phi = \dfrac{\pi}{3}$, and

β=0.5.

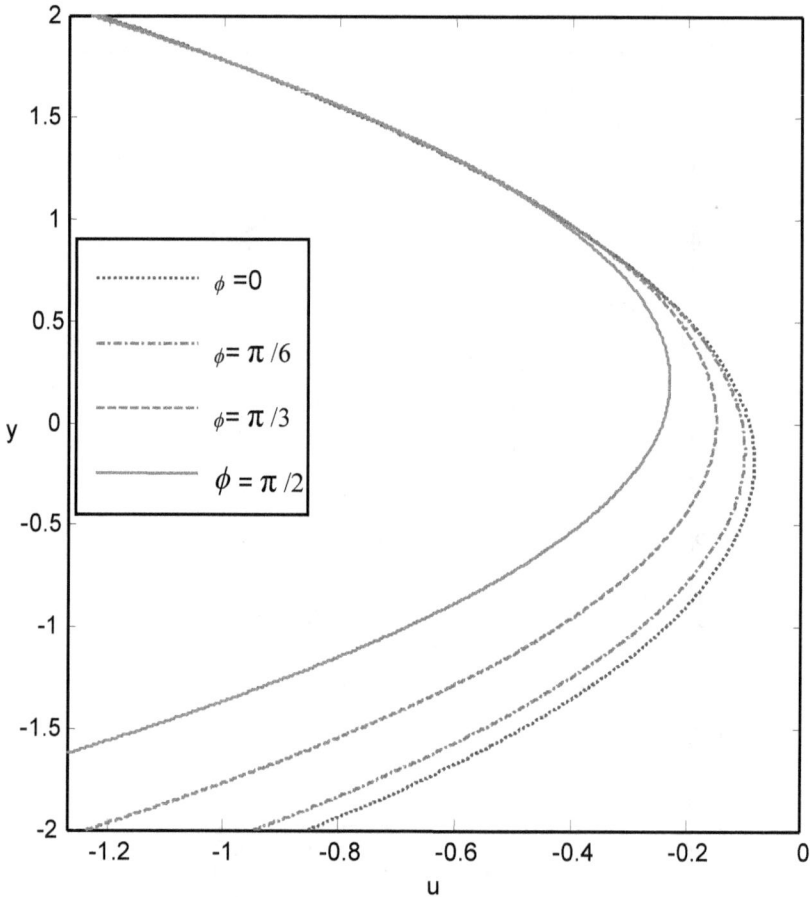

Figure 2.5: The velocity profiles with a=0.6, b=0.8, d=1.2, x=1, $\overline{Q}$=2, β=0.5 and We =0.01.

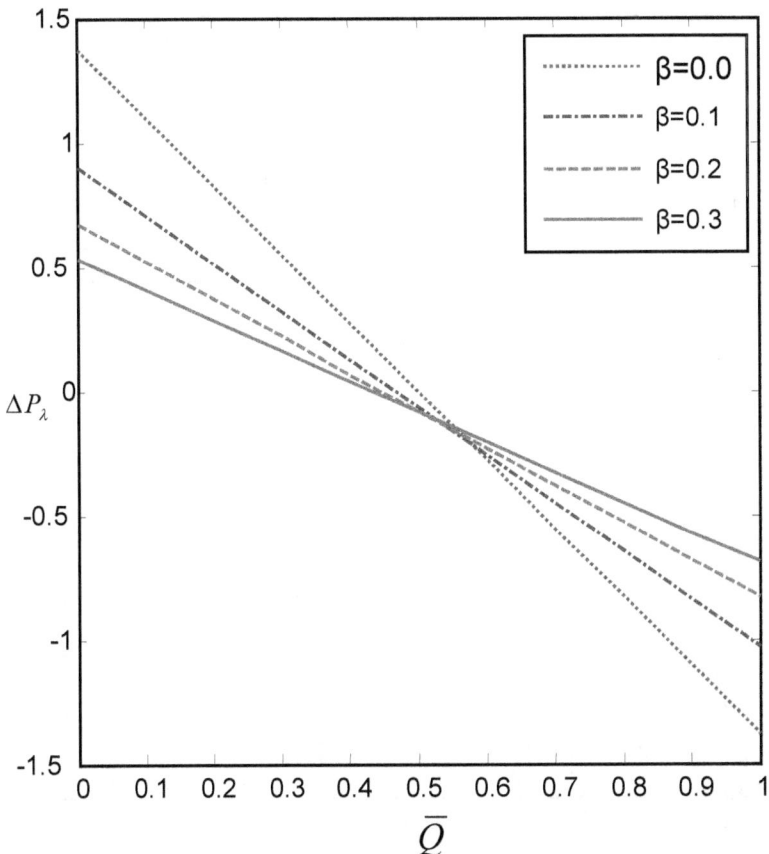

Figure 2.6: Various of $\bar{Q}$ with ΔP_λ for different values of permeability parameter including slip β with a=0.5, b=0.5, d=1, $\phi = \dfrac{\pi}{3}$ and We =0.01.

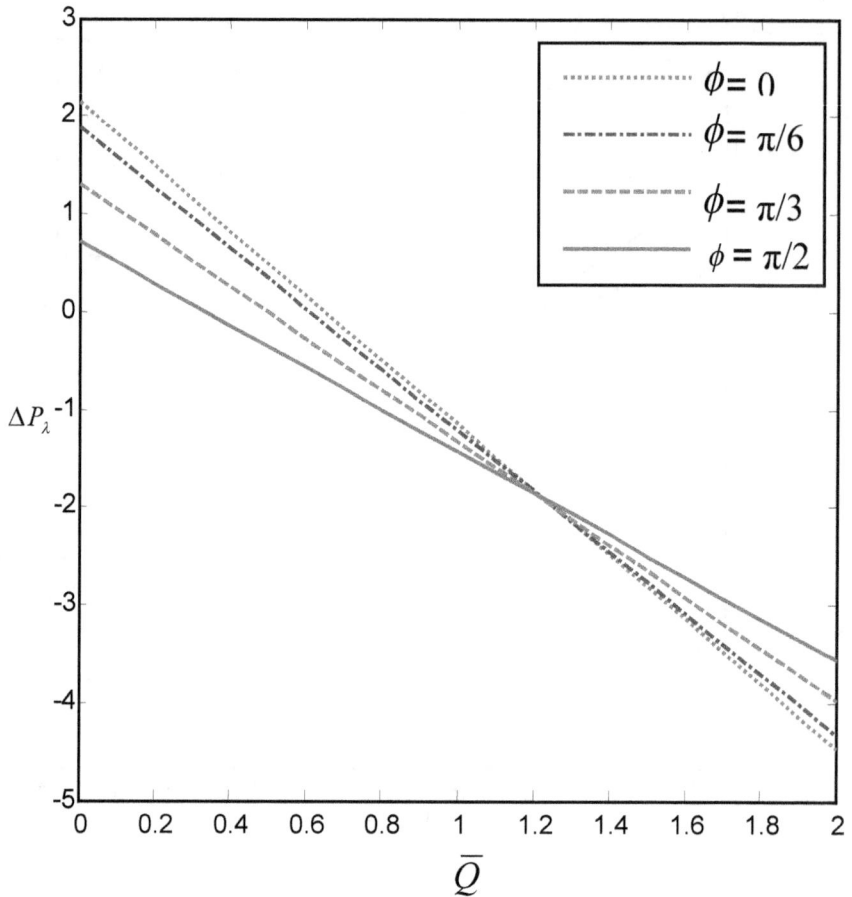

Figure 2.7: Various of $\overline{Q}$ with ΔP_λ for different values of phase difference ϕ with a=0.5, b=0.5, d=1, β=0.01 and We =0.01.

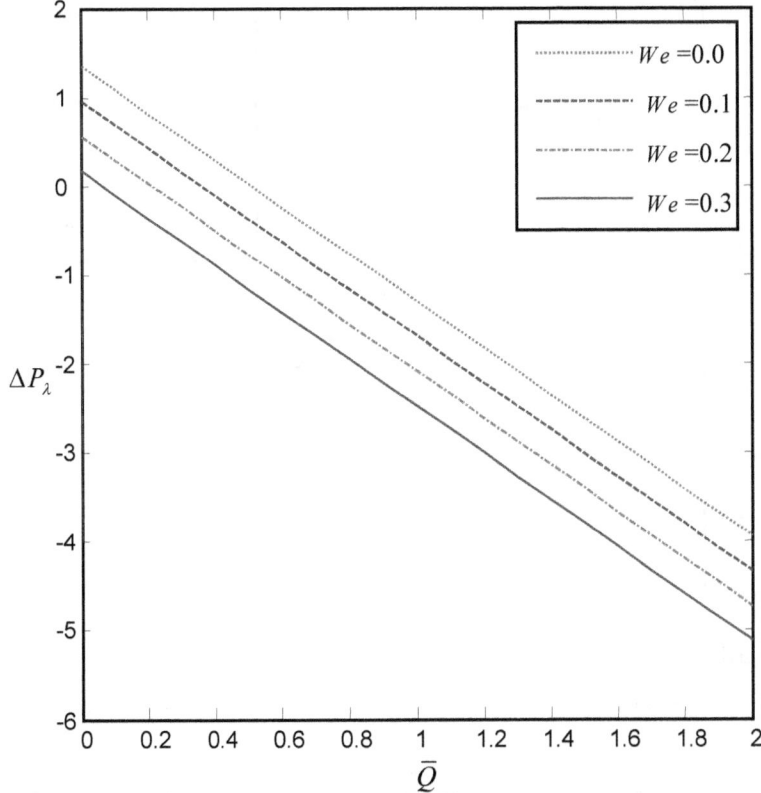

Figure 2.8: Various of $\overline{Q}$ with ΔP_λ for different values of Weissenberg number We with a=0.5, b=0.5, d=1, $\phi = \dfrac{\pi}{3}$ and β=0.01

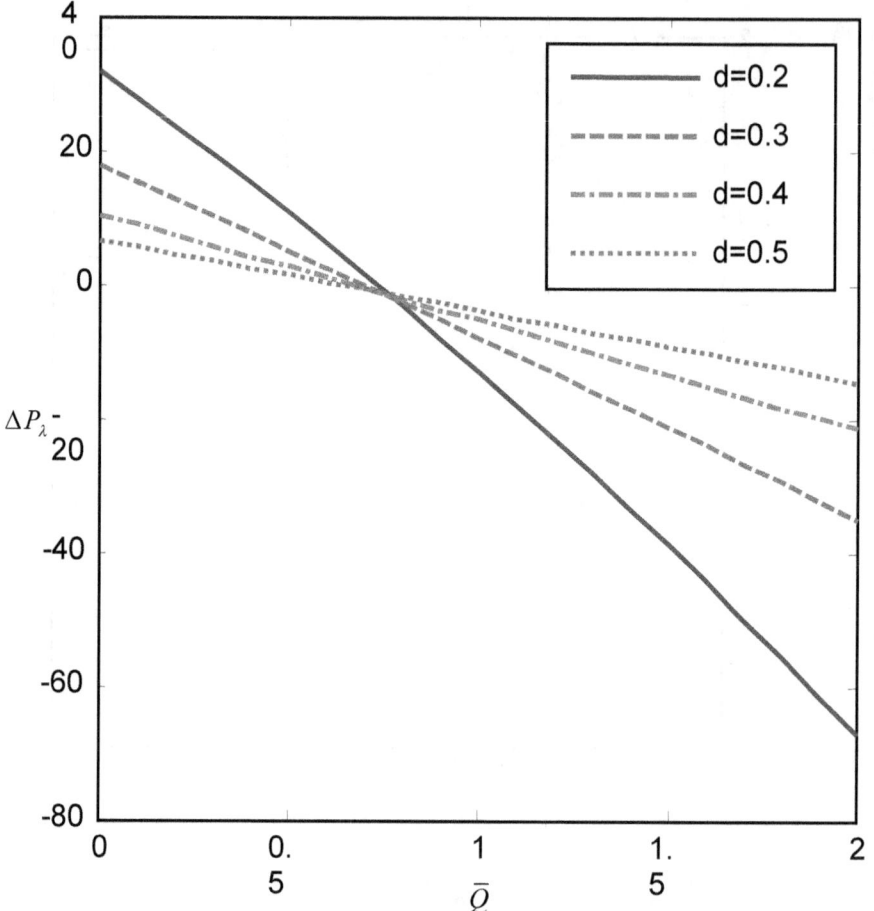

Figure 2.9: Various of $\bar{Q}$ with ΔP_λ for different values of channel width d with

a=0.5, b=0.5, $\phi = \dfrac{\pi}{3}$, β=0.01 and We =0.01.

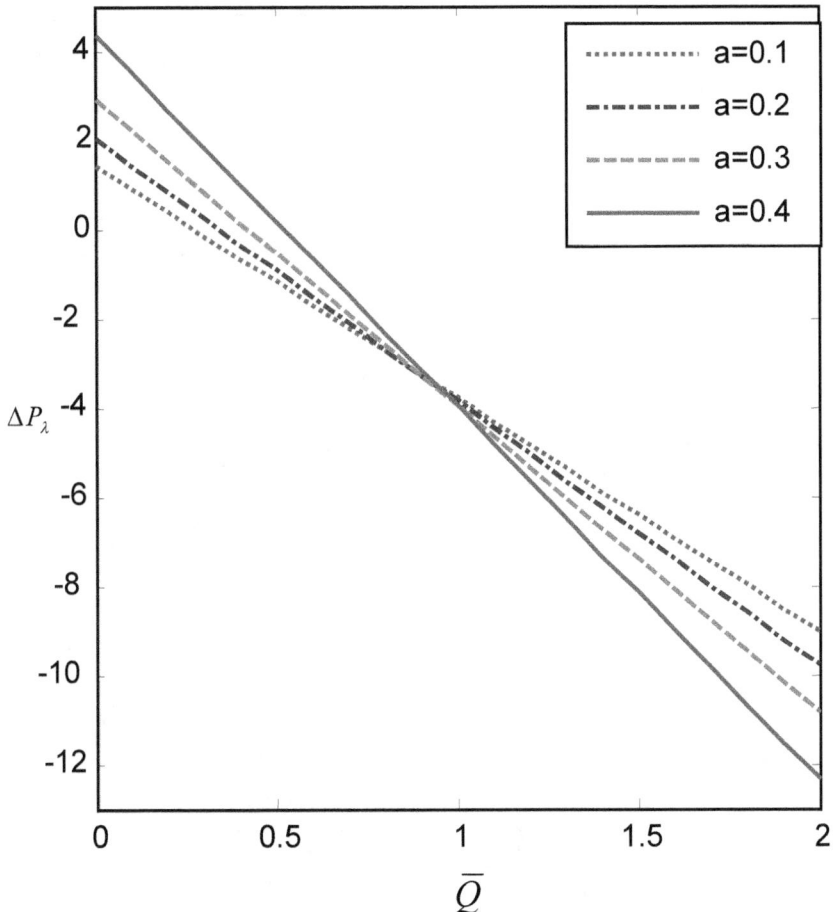

Figure 2.10: Various of $\overline{Q}$ with ΔP_λ for different values of **a** with   b=0.5,   d=1,

$\phi = \dfrac{\pi}{3}$, β=0.01 and We =0.01.

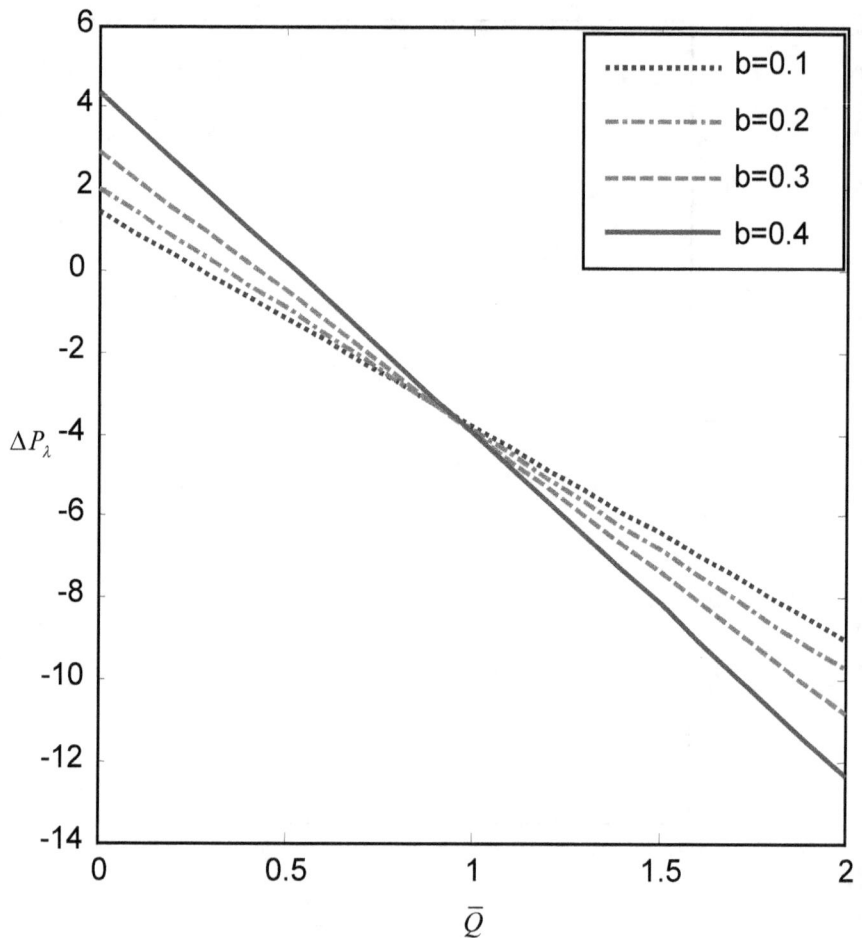

Figure 2.11: Various of $\overline{Q}$ with ΔP_λ for different vales of b with a=0.5, d=1, $\phi = \dfrac{\pi}{3}$, β=0.01 and We =0.01.

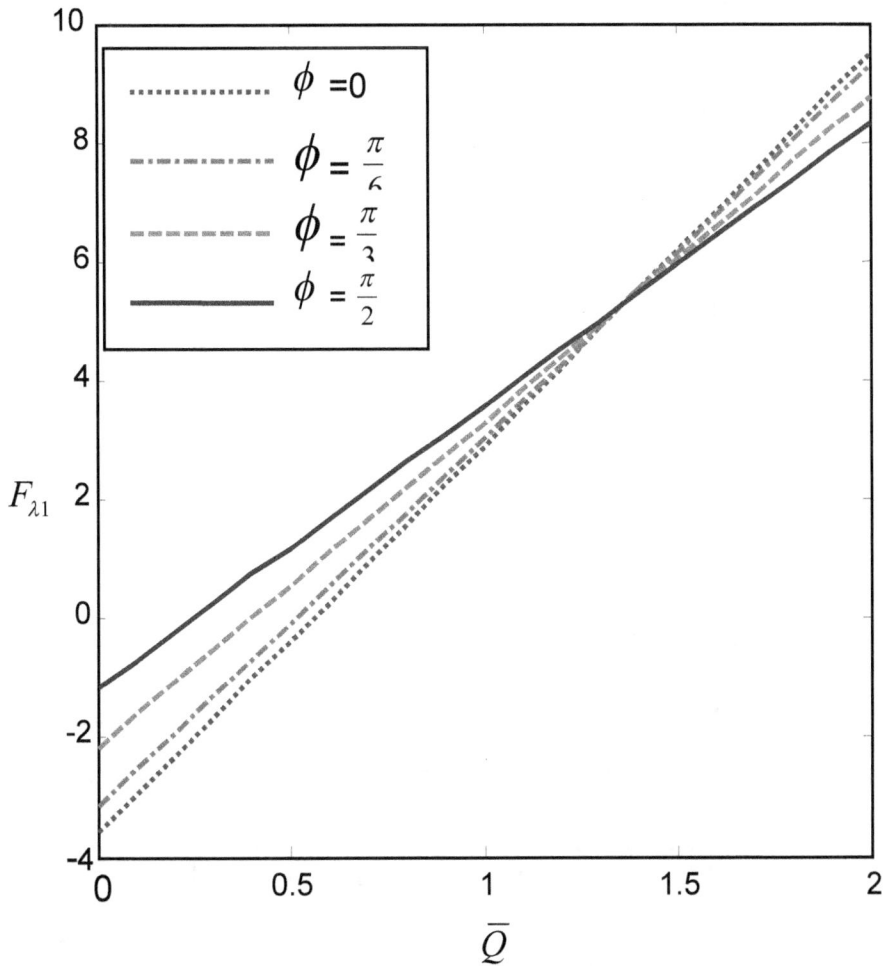

Figure 2.12: Various of frictional force $F_{\lambda 1}$ at y=h$_1$ at a=0.5, b=0.5, d=1, β=0.01

and We =0.01

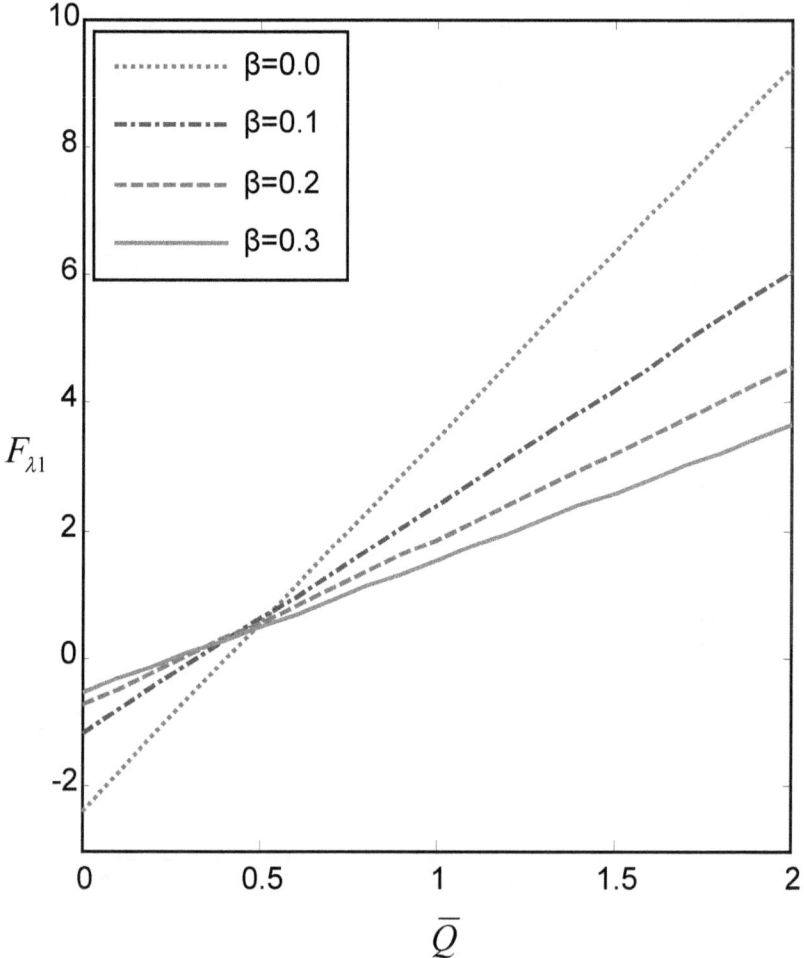

Figure 2.13: Various of frictional force $F_{\lambda 1}$ at y=h_1 at a=0.5, b=0.5, d=1, $\phi = \dfrac{\pi}{3}$ and

We =0.01.

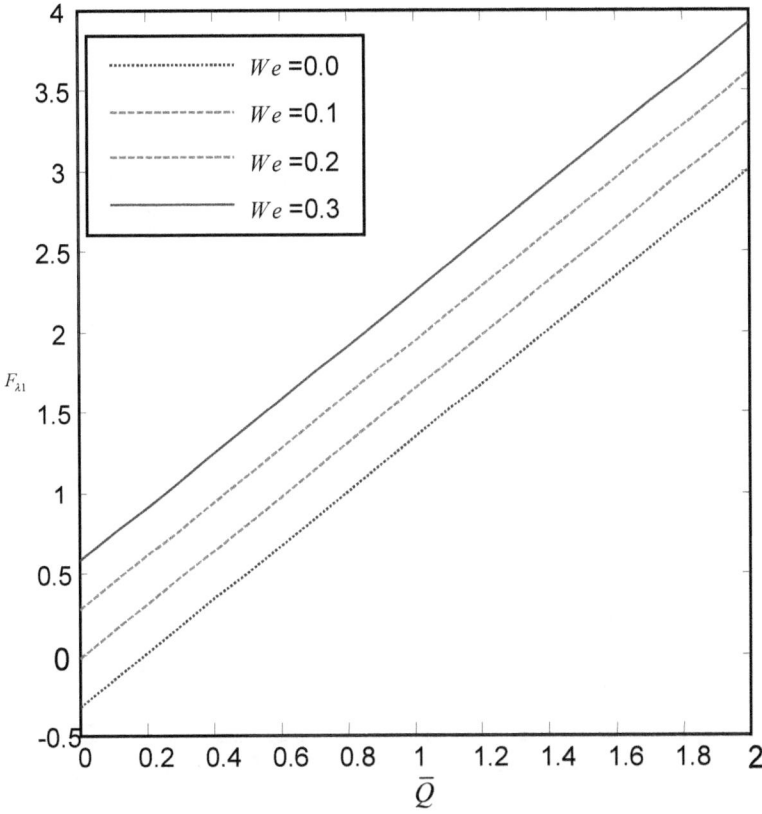

Figure 2.14: Various of frictional force $F_{\lambda 1}$ at y=h₁ at a=0.5, b=0.5, d=1, $\Phi = \dfrac{\pi}{3}$

and β=0.01.

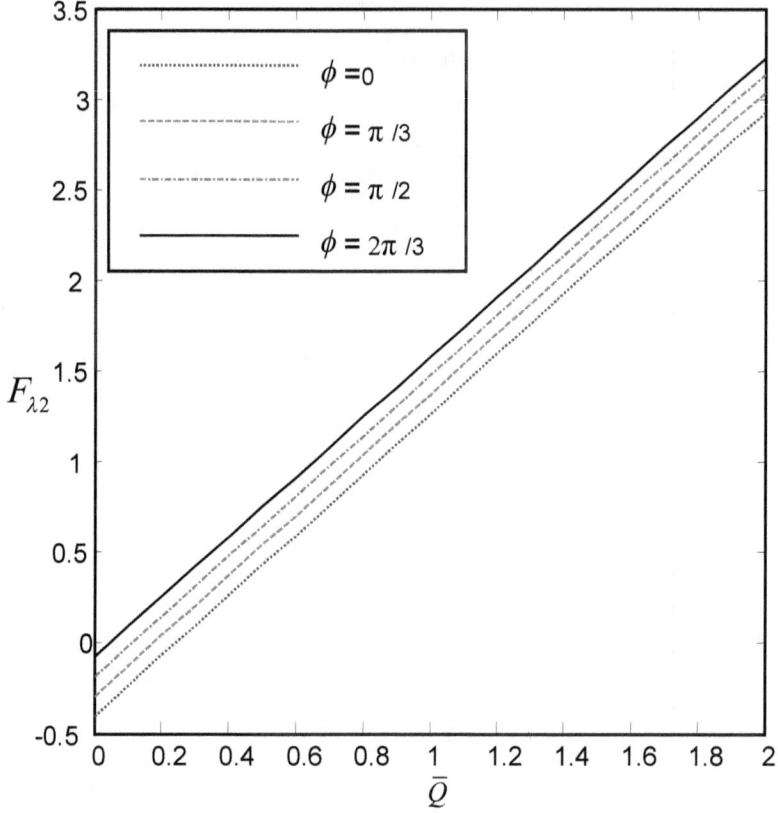

Figure 2.15: Various of frictional force $F_{\lambda 2}$ at y=h₂ at a=0.5, b=0.5, d=1, β=0.01 and We =0.01

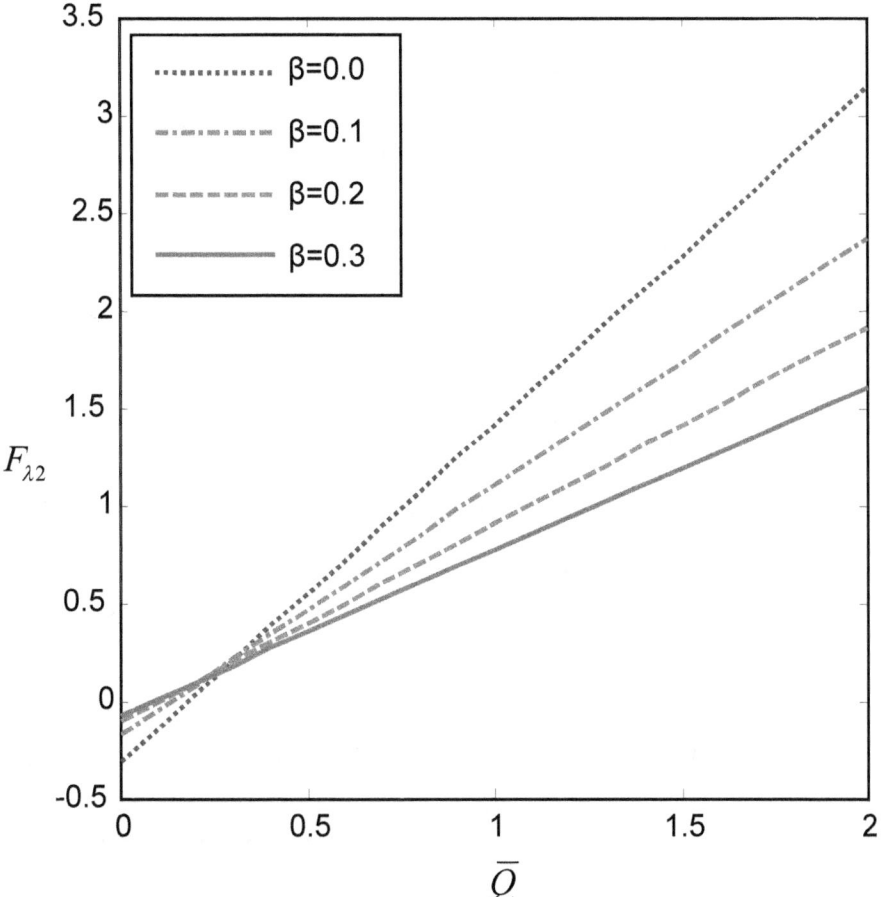

Figure 2.16: Various of frictional force $F_{\lambda 2}$ at y=h$_2$ with a=0.5, b=0.5, d=1, $\phi = \frac{\pi}{3}$,

β=0.01 and We =0.01

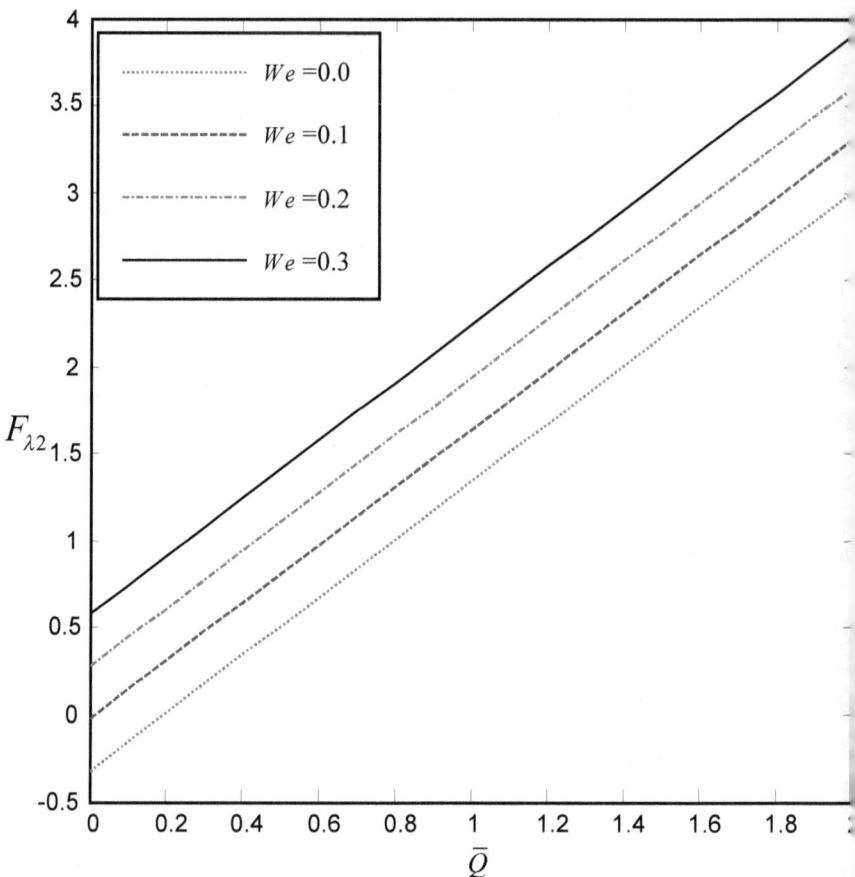

Figure 2.17: Various of frictional force $F_{\lambda 2}$ at y=h$_2$ with a=0.5, b=0.5, d=1, $\phi = \dfrac{\pi}{3}$ and β=0.01.

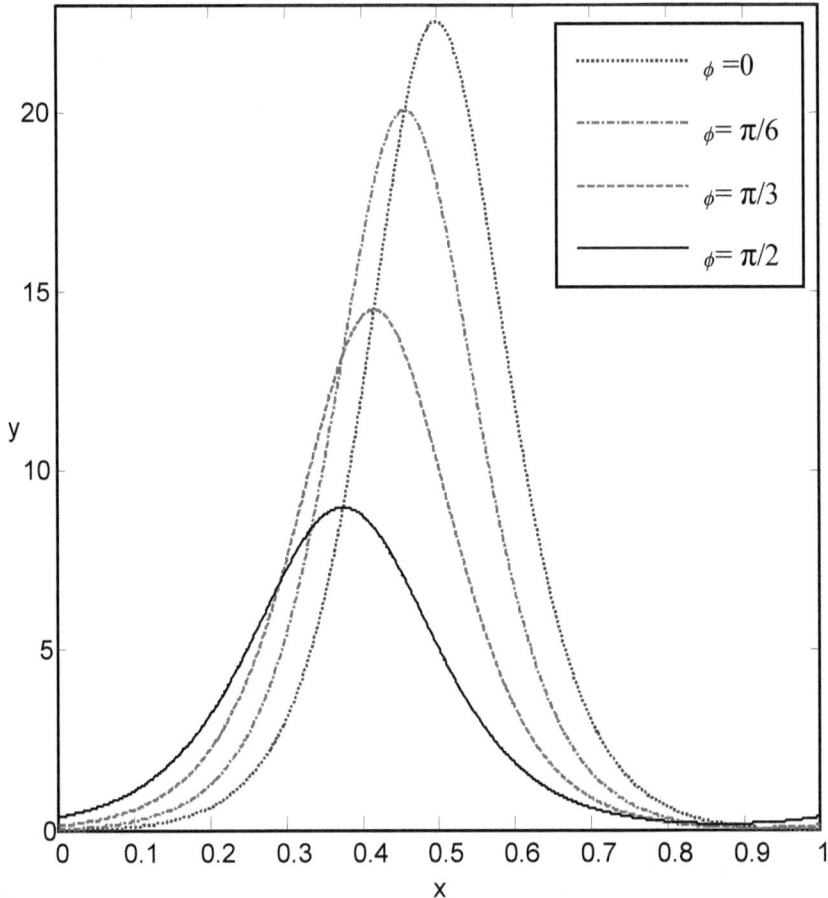

Figure 2.18: Various of pressure gradient $\dfrac{dp}{dx}$ with x for different value of phase

difference ϕ with a=0.5, b=0.5, d=1, $\overline{Q}$=-1, β=0.01 and We =0.01

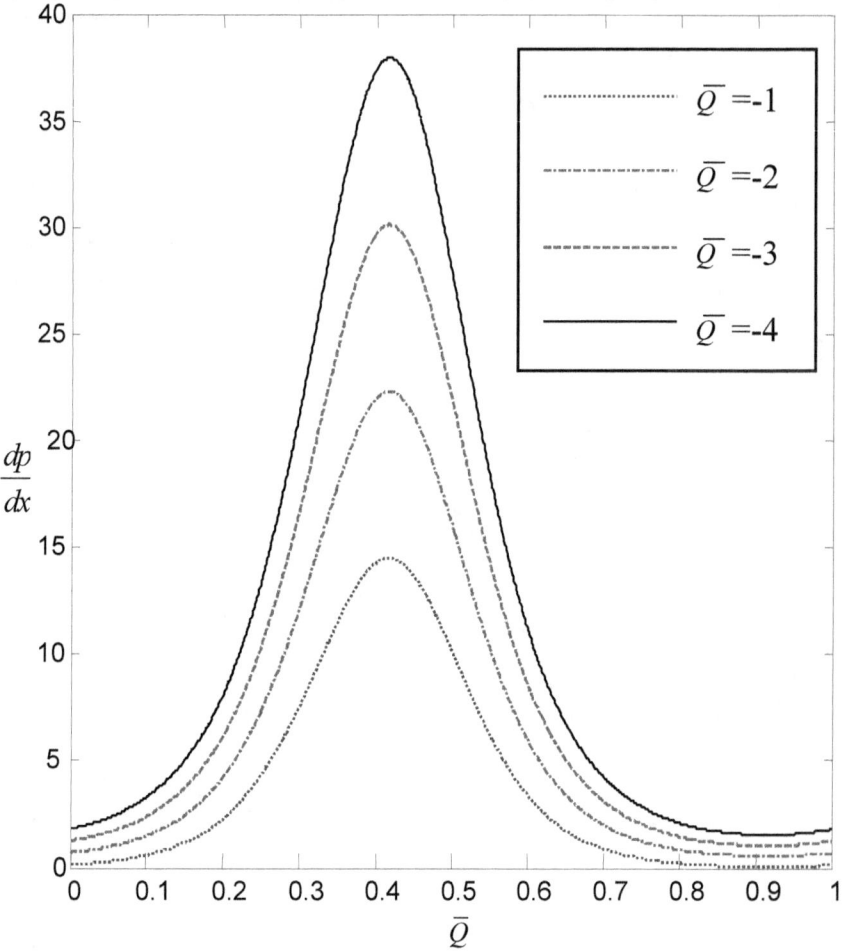

Figure 2.19: Various of pressure gradient $\dfrac{dp}{dx}$ with x for different value of average

flow rate $\overline{Q}$ with a=0.5, b=0.5, d=1, $\phi = \dfrac{\pi}{3}$, β=0.01 and We =0.01

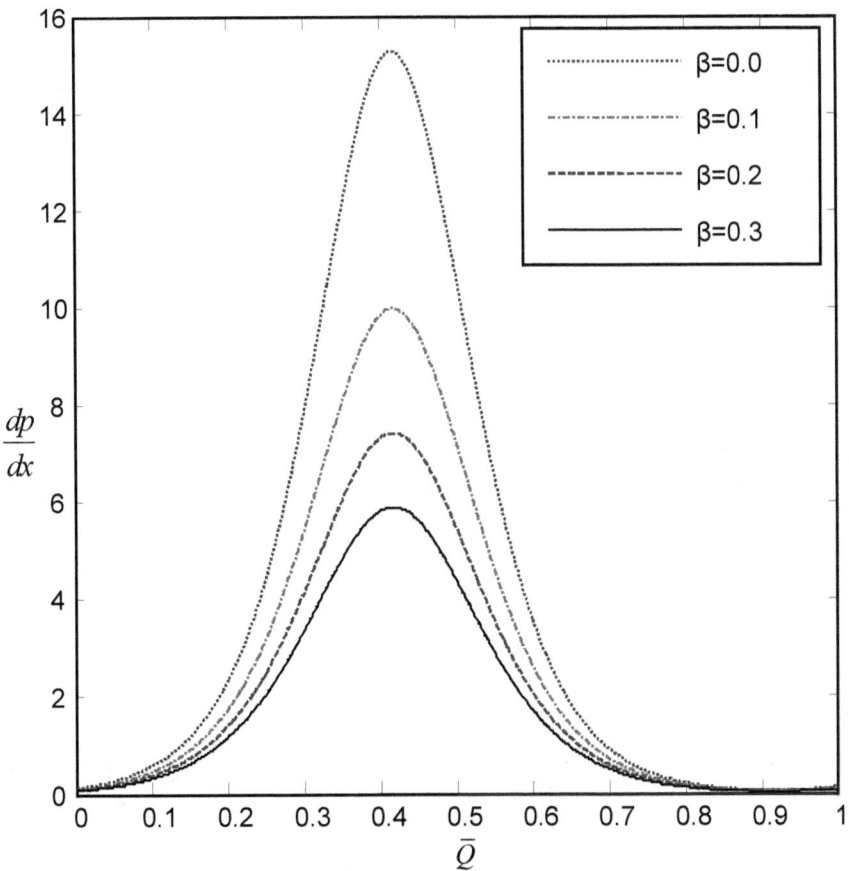

Figure 2.20: Various of pressure gradient $\dfrac{dp}{dx}$ with x for different value of permeability parameter β with a=0.5, b=0.5, d=1, $\overline{Q}$=-1, $\phi = \dfrac{\pi}{3}$ and We =0.01.

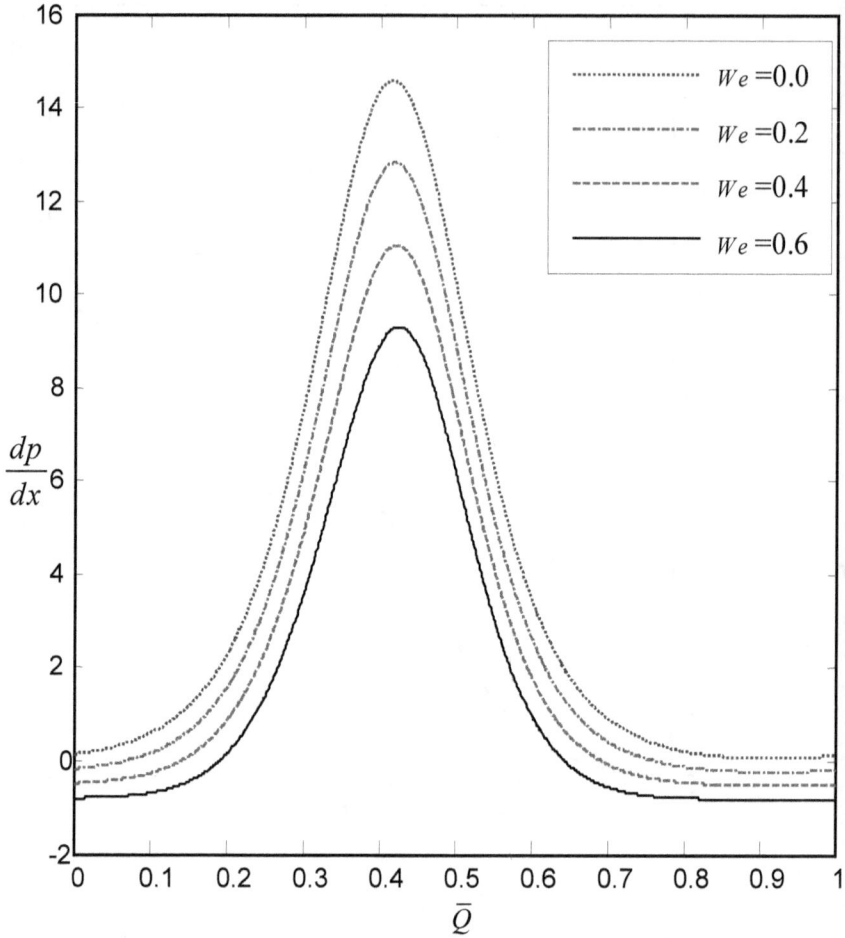

Figure 2.2: Various of pressure gradient $\dfrac{dp}{dx}$ with x for different value of

Weissenberg number We with a=0.5, b=0.5, d=1, $\overline{Q}$=-1, $\phi = \dfrac{\pi}{3}$ and

β=0.01.

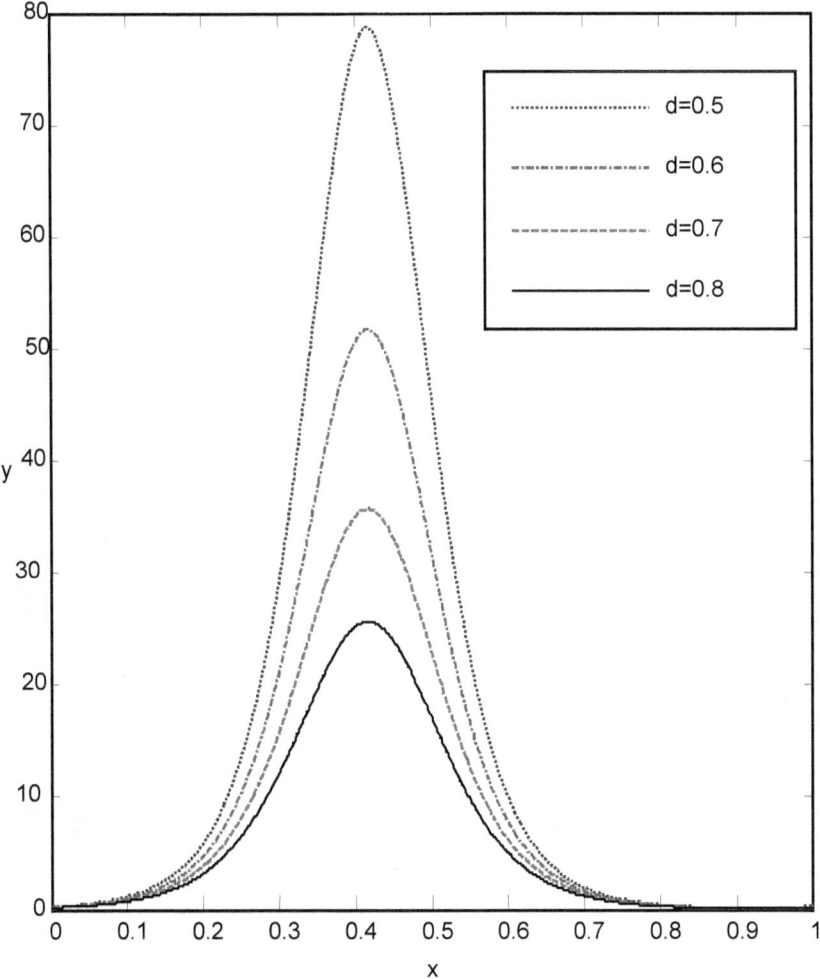

Figure 2.22: Various of pressure gradient $\dfrac{dp}{dx}$ with x for different value of channel

width d with a=0.5, b=0.5, $\overline{Q}$=-2, $\phi = \dfrac{\pi}{3}$, β=0.01 and We =0.01.

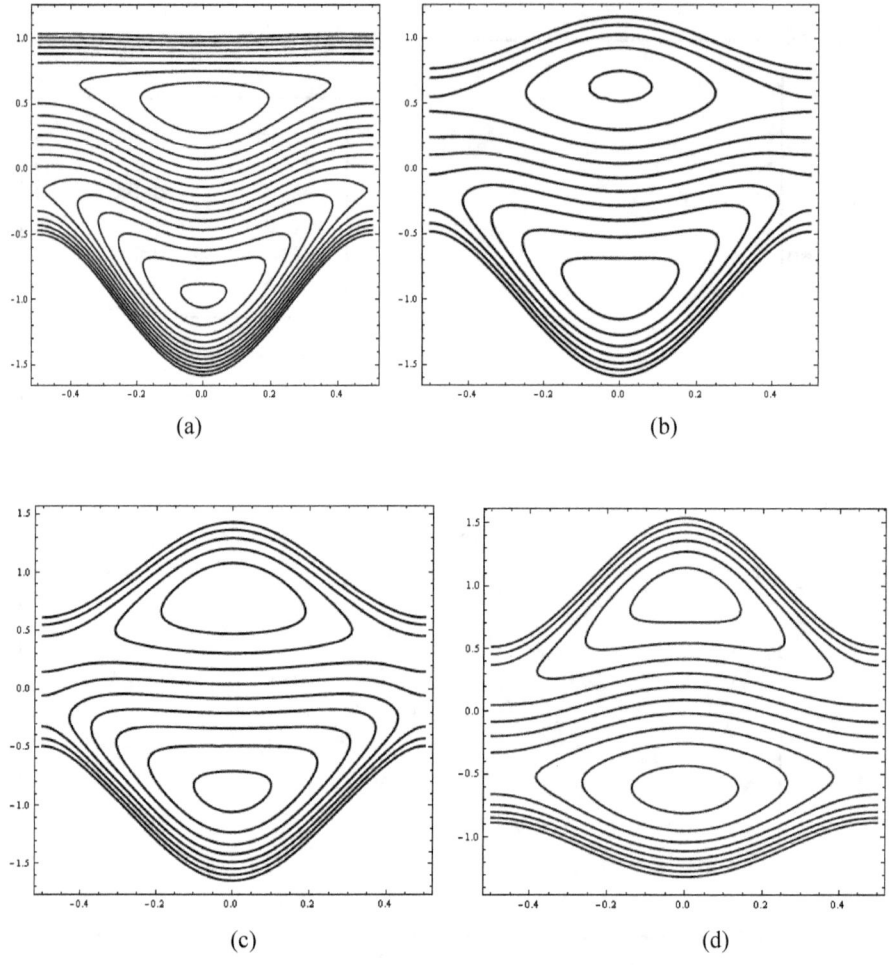

Figure 2.23: Streamlines in the wave frame for pumping with b=0.5, d=1, β=0.01, ϕ=0, $\overline{Q}$=2 and We =0.01, a) a=0, b) a=0.2, c) a=0.4 and d) a=0.6.

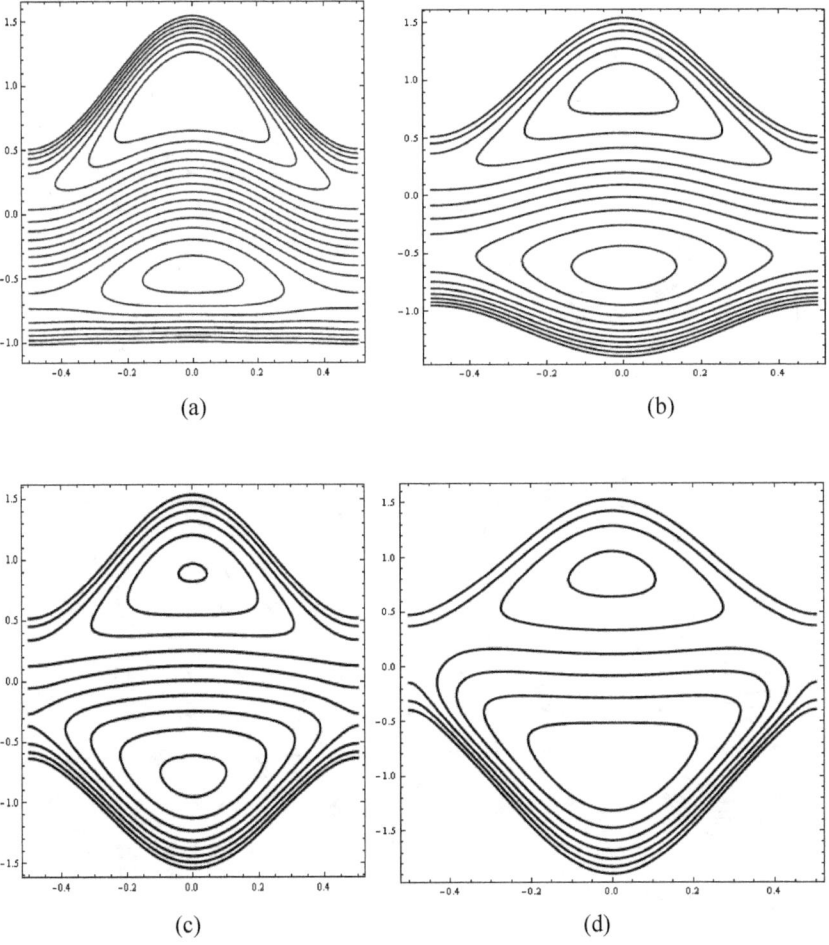

Figure 2.24: Streamlines in the wave frame for pumping with a=0.5, d=1, β=0.01, ϕ=0, $\overline{Q}$ =2 and We =0.01, a) b=0, b) b=0.2, c) b=0.4 and d) b=0.6.

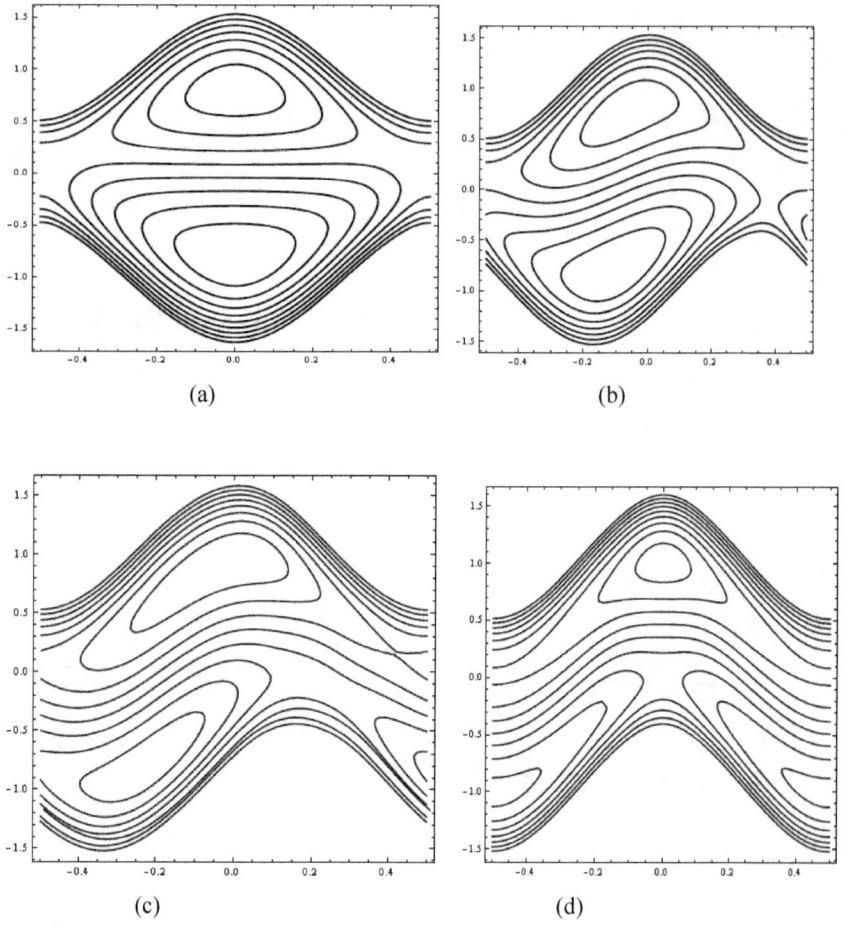

(a)   (b)

(c)   (d)

Figure 2.25: Streamlines in the wave frame for pumping with a=0.5, b=0.5, d=1, β=0.01, $\overline{Q}$=2, We =0.01, a) ϕ=0, b) ϕ= π /3, c) ϕ=2 π /3 and d) ϕ= π.

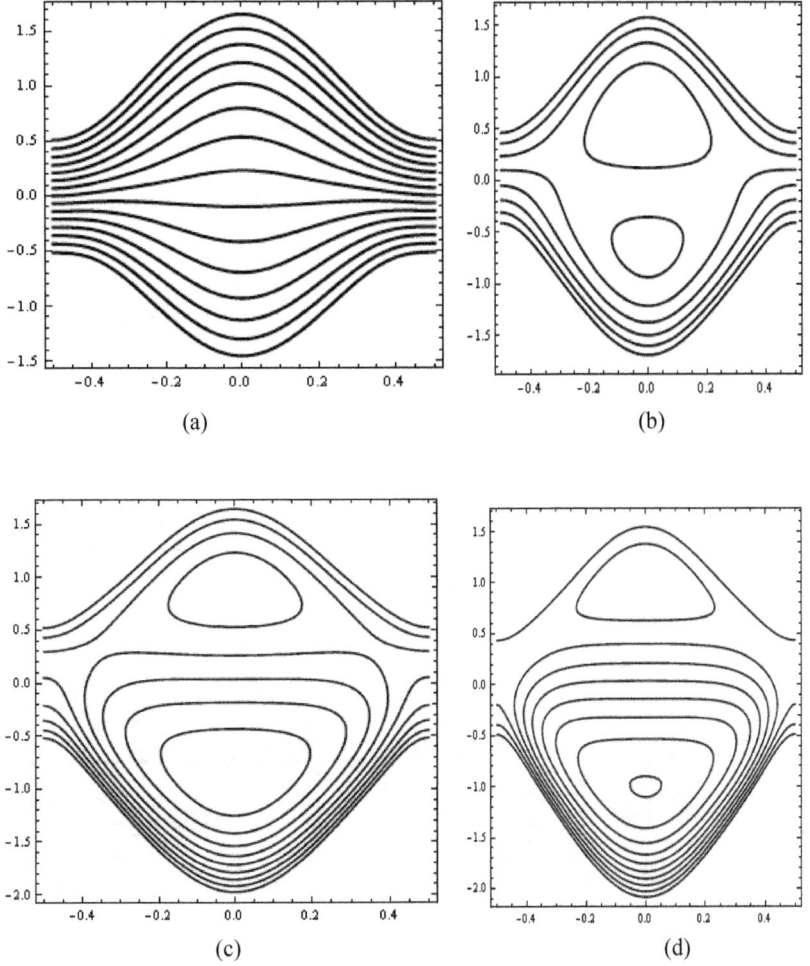

Figure 2.26: Streamlines in the wave frame for pumping with a=0.5, b=0.5, d=1,

ϕ=0, β=0.01 and We =0.01 a) $\overline{Q}$=1, b) $\overline{Q}$=1.5, c) $\overline{Q}$=2 and d) $\overline{Q}$=2.5.

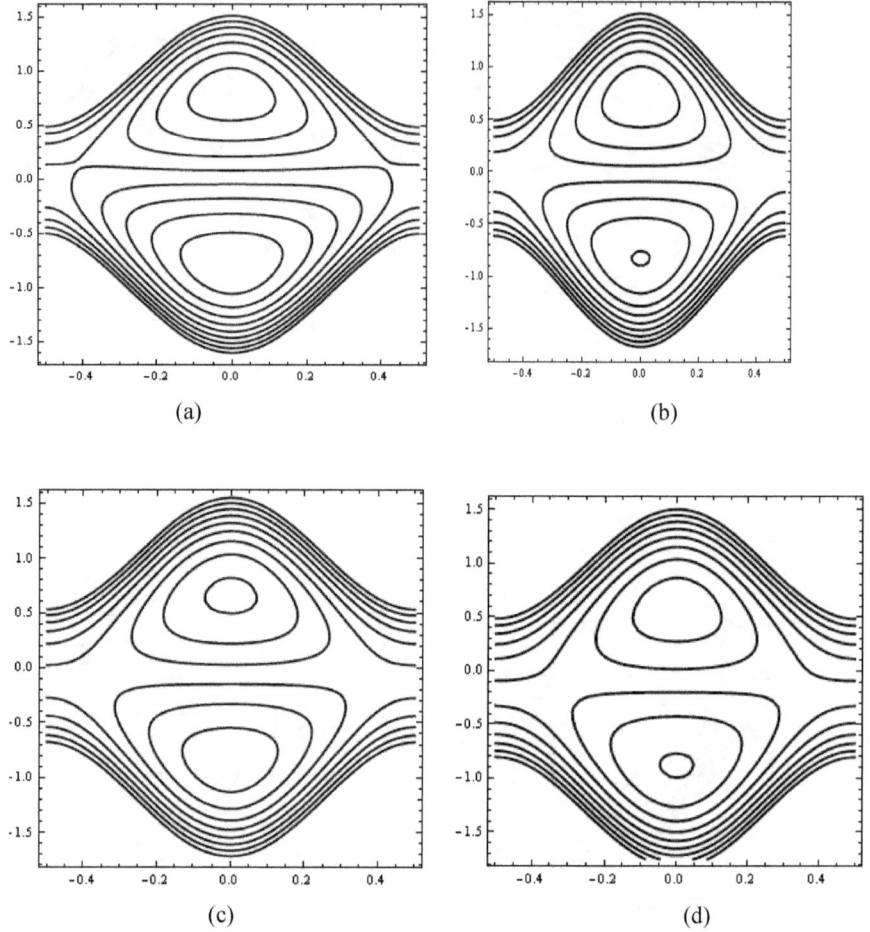

Figure 2.27: Streamlines in the wave frame for pumping with a=0.5, b=0.5, d=1,

ϕ=0, $\overline{Q}$=2, β=0.01 and We =0.01 a) d=1, b) d=1.1, c) d=1.2 and d) d=1.3.

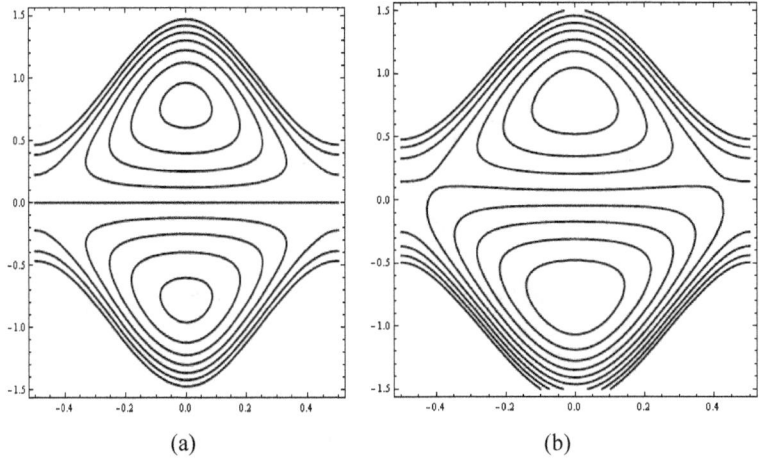

(a)   (b)

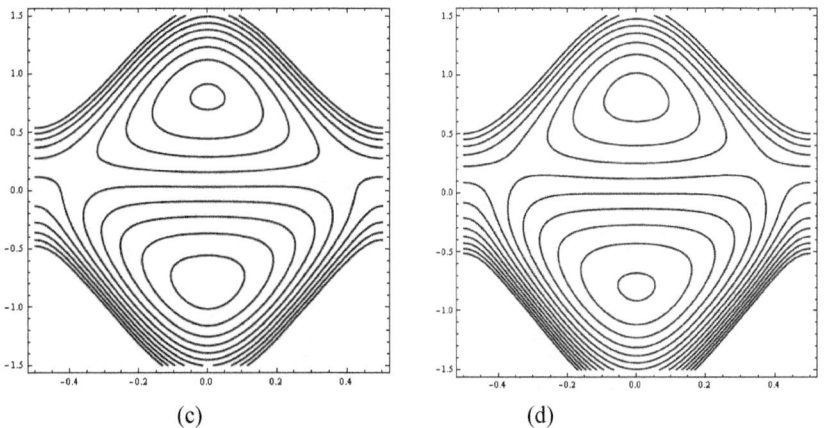

(c)   (d)

Figure 2.28: Streamlines in the wave frame for pumping with a=0.5, b=0.5, d=1,

ϕ=0, d=1, $\overline{Q}$=2 and β=0.01 a) We =0, b) We =0.01, c) We =0.02 and

d) We =0.03.

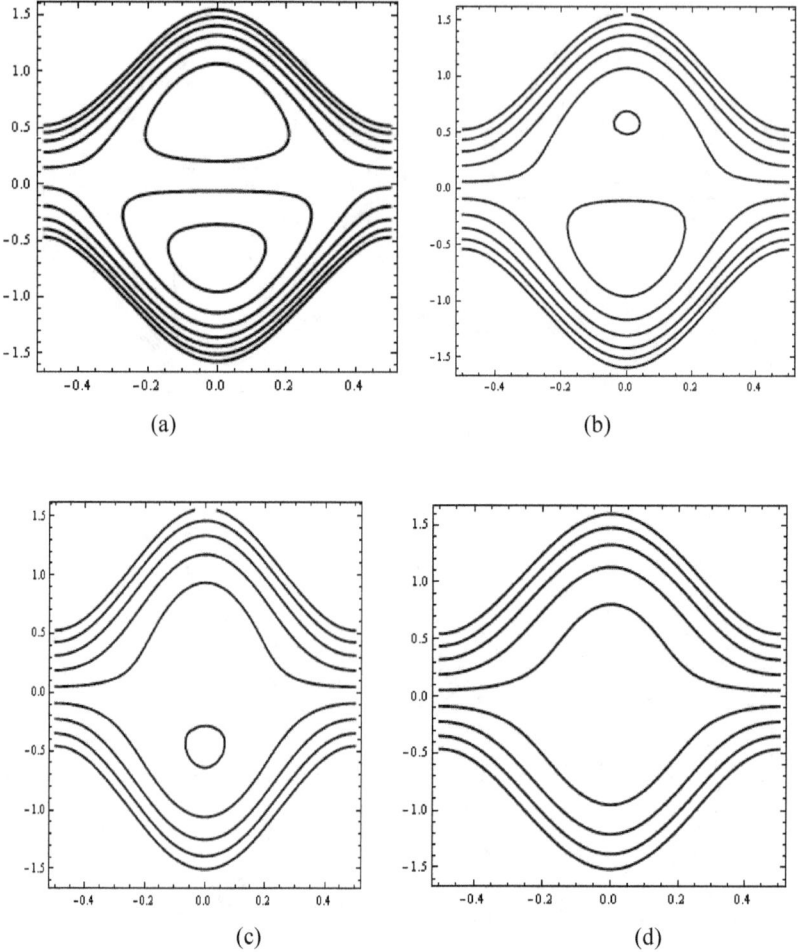

Figure 2.29: Streamlines in the wave frame for pumping with a=0.5, b=0.5, d=1,

ϕ=0, d=1, $\overline{Q}$ =2 and We =0.01 a) β=0, b) β=0.2, c) β=0.4 and d) β=0.6.

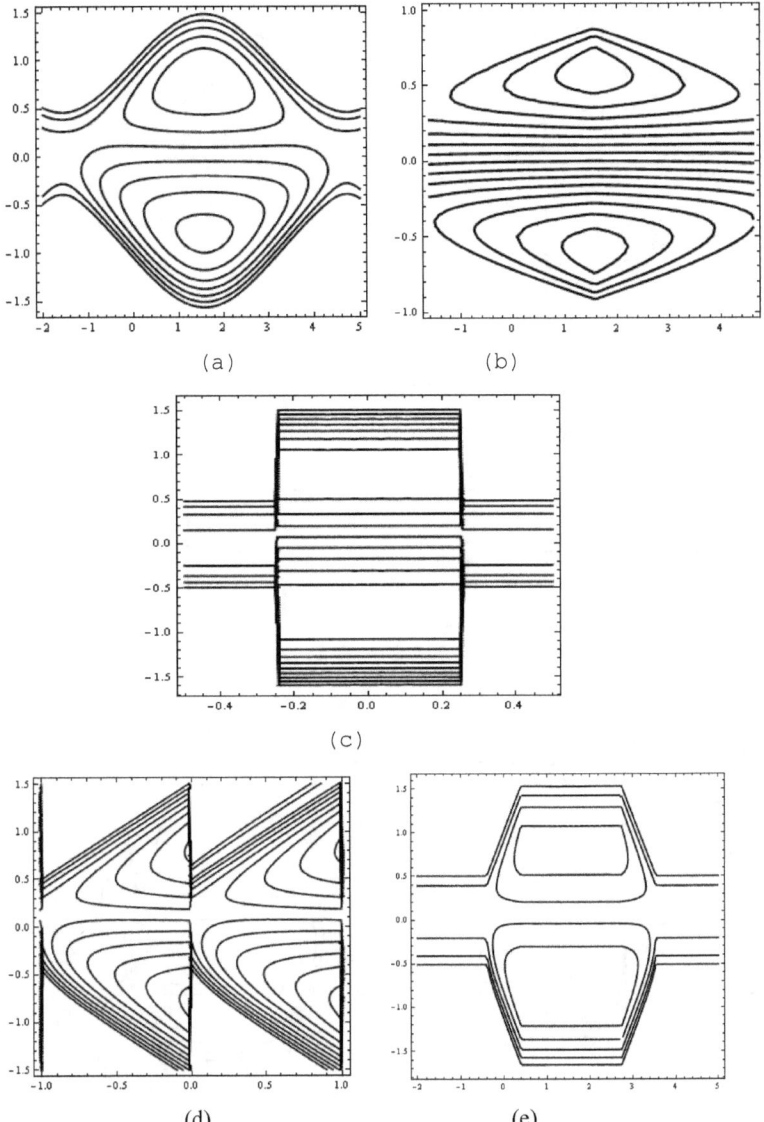

Figure 2.30 Streamlines for a=0.5, b=0.5, d=1, ϕ=0, d=1, $\overline{Q}$=2, β=0.4 and We =0.01 a) Sinusoidal wave, b) Triangular wave, c) Square wave, d) Sawtooth wave and e) Trapezoidal wave.

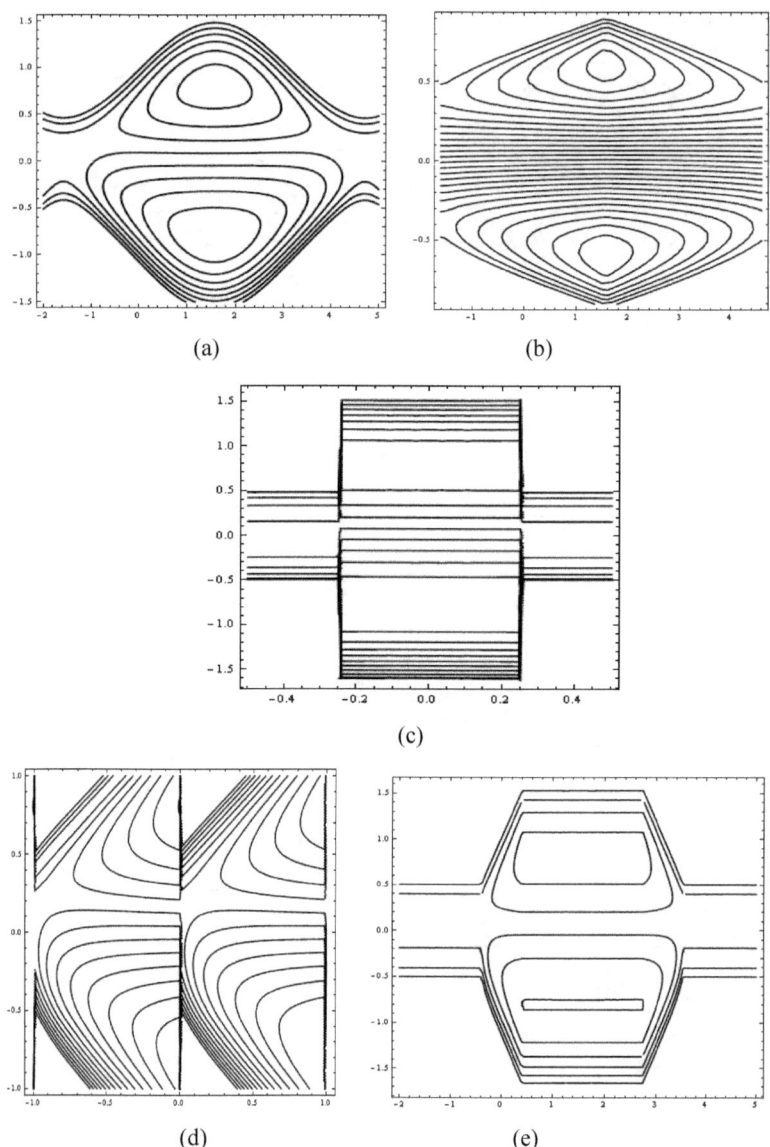

Figure 2.31: Streamlines for a=0.5, b=0.5, d=1, ϕ=0, d=1, $\overline{Q}$=2, β=0.0 and We =0.01 a) Sinusoidal wave, b) Triangular wave, c) Square wave, d) Sawtooth wave and e) Trapezoidal wave.

3.1 Introduction

The study of peristaltic transport in fluid mechanics has been carried out by many researchers to look for its implication in the biological sciences in general and in biomechanics, in particular. The peristaltic flow occurs due to the contraction and expansion of a progressive wave propagating along the length of a distensible tube containing fluid. Peristaltic wave causing transportation of the fluid through muscular tube is indeed an important biological mechanism responsible for various physiological functions of the organs of the human body. Such mechanism may be involved in urine transport from kidney to bladder through the ureter, swallowing foods though the esophagus, the transport of spermatozoa in the cervical canal, the movement of chyme in small intestines and in the transport of bile. Such a wide occurrence of peristaltic motion should not be surprising at all since it results physiologically from neuro-muscular properties of any tubular smooth muscle.

Most of the previous works in the literature deals with the peristaltic flow in a symmetric channel or tube. Due attention has not been given to the peristaltic mechanism in an asymmetric channel. Recently, physiologists observed that myometrial contractions may occur in both symmetric and asymmetric directions. Eytan and Elad (1999) have presented a mathematical model of wall induced peristaltic fluid flow in a two-dimensional channel with wave trains having a phase difference moving independently on the upper and lower walls to stimulate intra-uterine fluid motion in a sagittal cross –section of the uterus. In another paper, Eytan *et al.* (1999) noted that the width of the sagittal cross-section of the uterine cavity increases towards the fundus and the cavity is not fully occluded during the contractions. Mishra and Rao (2004) discussed the peristaltic motion of viscous fluid in a two-dimensional asymmetric channel under long wavelength assumption. Rao and Mishra (2004) also analyzed the curvature effects on peristaltic transport in an asymmetric channel. They obtained the perturbation solution when ratio of the channel width to the wavelength is small. Recenty Ali and Hayat (2007) made a detailed study as the peristaltic motion of a Carreau fluid in an asymmetric channel with impermeable walls. Since many biological systems such as blood vessels contion a tissue layer it will be interesting to study peristaltic transport of a biofluid through asymmetric channel with permeable walls.

The main objective of present investigation is to put forward the analysis of peristaltic flow of a non-Newtonian fluid in an asymmetric channel. For that we present rheological constitutive equations of Carreau fluid. The Carreau fluid model is a four parameter model and has useful properties of a truncated power law model that does not have a discontinuous first derivative. It possesses shear thinning i.e. the viscosity decreases by increasing shear rate, some times reaching to 10^{-3} or 10^{-4} for a zero shear rate. The relevant equations for the fluid under consideration have been first modeled and then solved. The assumption for the solution is that wavelength of the peristaltic wave is long. A regular perturbation technique is employed to solve the present problem and solutions are expanded in a power of small Weissenberg number. The analysis is made for the stream function, axial pressure gradient and pressure drop over a wavelength. The influence of emerging parameters is shown on pumping, pressure gradient and trapping.

3.2 Mathematical formulation

Let us consider the peristaltic transport of an incompressible Carreau fluid in a two–dimensional channel of width $d_1 + d_2$ (Figure 3.1). The flow is generated by sinusoidal wave trains propagating with constant speed c along the permeable walls of the channel. The geometry of the wall surfaces is defined as.

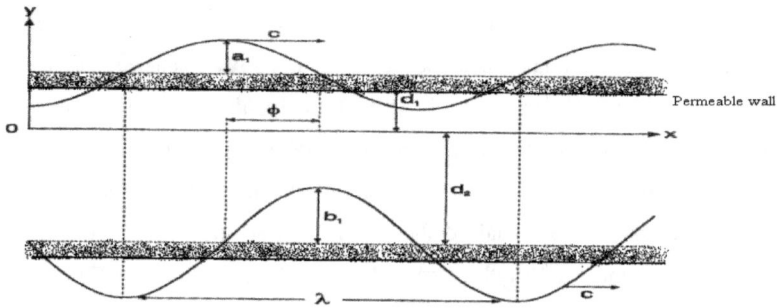

Figure 3.1.Schematic diagram of a two-dimensional asymmetric channel

$$h_1\left(\overline{X},\overline{t}\right)=\overline{d}_1+\overline{a}_1\cos\left[\frac{2\pi}{\lambda}\left(\overline{X}-ct\right)\right]\ \dots\quad\text{upper wall,}\qquad(3.1)$$

$$h_2\left(\overline{X},\overline{t}\right)=-\overline{d}_2-\overline{b}_1\cos\left[\frac{2\pi}{\lambda}\left(\overline{X}-ct\right)+\phi\right]\dots\dots\quad\text{lower wall.}\qquad(3.2)$$

in which $\overline{a}_1$ and $\overline{b}_1$ are the amplitudes of the waves, λ is the wave length, c is the wave speed, ϕ $(0\le\phi\le\pi)$ is the phase difference, $\overline{X}$ measured along the axis of the channel and $\overline{Y}$ perpendicular to $\overline{X}$. Let $\left(\overline{U},\overline{V}\right)$ be the velocity components in fixed frame of reference $\left(\overline{X},\overline{Y}\right)$. It should be noted that $\phi=0$ corresponding to symmetric channel with waves our of phase and for $\phi=\pi$ the waves are in phase, and further $\overline{a}_1,\ \overline{a}_2,\ \overline{d}_1,\ \overline{d}_2$ and ϕ satisfies the condition

$$\overline{a}_1^{\ 2}+\overline{b}_1^{\ 2}+2\overline{a}_1\overline{b}_1\cos\phi\le\left(\overline{d}_1+\overline{d}_2\right)^2\qquad.$$

(3.3)

3.3 Equations of motion

The constitutive equation for a Carreau fluid is

$$\overline{\tau}=-\left[\eta_\infty+\left(\eta_0-\eta_\infty\right)\!\left(1+\left(\Gamma\overline{\dot{\gamma}}\right)^2\right)^{\frac{n-1}{2}}\right]\overline{\dot{\gamma}},\qquad(3.4)$$

where $\overline{\tau}$ is the extra stress tensor, η_∞ is the infinite shear – rate viscosity, η_0 is the zero shear-rate viscosity, Γ is the time constant, n is the dimensionless power law index and $\dot{\gamma}$ is defined as

$$\overline{\dot{\gamma}}=\sqrt{\frac{1}{2}\sum_i\sum_j\overline{\dot{\gamma}}_{ij}\overline{\dot{\gamma}}_{ji}}=\sqrt{\frac{1}{2}\Pi}\qquad.$$

(3.5)

here Π is the second invariant of strain- rate tensor. We consider in the constitutive equation (3.4) the case for which $\eta_\infty=0$, and so we can write

$$\overline{\tau}=-\eta_0\left[1+\left(\Gamma\overline{\dot{\gamma}}\right)\right]^{\frac{n-1}{2}}\overline{\dot{\gamma}}.\qquad(3.6)$$

Note that the above model reduces to Newtonian Model for $n=1$ or $\Gamma=0$

In the laboratory frame $(\overline{X},\overline{Y})$ the flow is unsteady. However, if observed in a coordinate system moving at the wave speed c (wave frame) $(\overline{x}, \overline{y})$ it can be treated as steady. The coordinates and velocities in the two frames are

$$\overline{x} = \overline{X} - c\overline{t}, \quad \overline{y} = \overline{Y}, \ \overline{u}\,(\overline{x},\overline{y}) = \overline{U} - c, \ \overline{v}\,(\overline{x},\overline{y}) = \overline{V}, \tag{3.7}$$

where $\overline{u}$ and $\overline{v}$ indicate the velocity components in the wave frame.

The incompressibility condition and scalar form of equation of motion give

$$\frac{\partial \overline{u}}{\partial \overline{x}} + \frac{\partial \overline{v}}{\partial \overline{y}} = 0, \tag{3.8}$$

$$\rho\left(\overline{u}\,\frac{\partial}{\partial \overline{x}} + \overline{v}\,\frac{\partial}{\partial \overline{y}}\right)\overline{u} = -\frac{\partial \overline{p}}{\partial \overline{x}} + \frac{\partial \overline{\tau}_{\overline{xx}}}{\partial \overline{x}} + \frac{\partial \overline{\tau}_{\overline{xy}}}{\partial \overline{y}},$$

(3.9a)

$$\rho\left(\overline{u}\,\frac{\partial}{\partial \overline{x}} + \overline{v}\,\frac{\partial}{\partial \overline{y}}\right)\overline{v} = -\frac{\partial \overline{p}}{\partial \overline{y}} + \frac{\partial \overline{\tau}_{\overline{xy}}}{\partial \overline{x}} + \frac{\partial \overline{\tau}_{\overline{xx}}}{\partial \overline{y}}.$$

(3.9b)

The flow in the permeable wall is described by Darcy's law. The velocity in the permeable wall is given by

$$\overline{u}_D = -\frac{k}{\eta_0}\frac{\partial \overline{p}}{\partial \overline{x}}. \tag{3.10}$$

Employing the transformations in above equations, introducing

$$x = \frac{\overline{x}}{\lambda}, \ y = \frac{\overline{y}}{d_1}, \ u = \frac{\overline{u}}{c}, \ v = \frac{\overline{v}}{c}, \ t = \frac{c}{\lambda}\overline{t}, \ h_1 = \frac{\overline{h_1}}{d_1}, u_D = \frac{\overline{u}_D}{c},$$

$$h_2 = \frac{\overline{h_2}}{d_1}, \ \tau_{xx} = \frac{\lambda}{\eta_0 c}\overline{\tau}_{\overline{xy}}, \ \tau_{yy} = \frac{d_1}{\eta_0 c}\overline{\tau}_{\overline{yy}}, \ \delta = \frac{d_1}{\lambda}, Re = \frac{\rho c \overline{d_1}}{\eta_0}, \tag{3.11}$$

$$We = \frac{\Gamma c}{d_1}, \quad p = \frac{\overline{d_1}^2}{c\lambda\eta_0}\overline{p}, \quad \dot{\gamma} = \frac{\overline{\dot{\gamma}}\,d_1}{c}, \quad a = \frac{\overline{a_1}}{d_1}, \quad b = \frac{\overline{a_2}}{d_1}, \quad d = \frac{\overline{d_2}}{d_1} \text{ and}$$

$$Da = \frac{k}{d_1^2}$$

and defining the stream function $\psi\,(x, y)$ by

$$u = \frac{\partial \psi}{\partial y}, \qquad v = -\delta \frac{\partial \psi}{\partial x}. \tag{3.12}$$

we find that equation (3.8) is identically satisfied and equations (3.9a), (3.9b) and (3.10) become

$$\delta \operatorname{Re}\left[\left(\frac{\partial \psi}{\partial y}\frac{\partial}{\partial x} - \frac{\partial \psi}{\partial x}\frac{\partial}{\partial y}\right)\frac{\partial \psi}{\partial y}\right] = -\frac{\partial p}{\partial x} + \frac{\delta^2}{2}\frac{\partial \tau_{xx}}{\partial x} + \frac{\partial \tau_{xy}}{\partial y},$$

(3.13a)

$$\delta^3 \operatorname{Re}\left[\left(\frac{\partial \psi}{\partial y}\frac{\partial}{\partial x} - \frac{\partial \psi}{\partial x}\frac{\partial}{\partial y}\right)\frac{\partial \psi}{\partial x}\right] = -\frac{\partial p}{\partial y} + \delta^2 \frac{\partial \tau_{xx}}{\partial x} + \delta \frac{\partial \tau_{xy}}{\partial y}. \tag{3.13b}$$

$$u_D = -Da\frac{\partial p}{\partial x}. \tag{3.14}$$

where

$$\tau_{xx} = 2\left[1 + \frac{(n-1)}{2}We^2 \dot{\gamma}^2\right]\frac{\partial^2 \psi}{\partial x \partial y}, \tag{3.15}$$

$$\tau_{xy} = \left[1 + \frac{(n-1)}{2}We^2 \dot{\gamma}^2\right]\left(\frac{\partial^2 \psi}{\partial y^2} - \delta^2 \frac{\partial^2 \psi}{\partial x^2}\right), \tag{3.16}$$

$$\tau_{yy} = -2\delta\left[1 + \frac{(n-1)}{2}We^2 \dot{\gamma}^2\right]\frac{\partial^2 \psi}{\partial x \partial y}, \tag{3.17}$$

$$\dot{\gamma} = \left[2\delta^2\left(\frac{\partial^2 \psi}{\partial x \partial y}\right)^2 + \left(\frac{\partial^2 \psi}{\partial y^2} - \delta^2 \frac{\partial^2 \psi}{\partial x^2}\right) + 2\delta^2\left(\frac{\partial^2 \psi}{\partial x \partial y}\right)^2\right]^{\frac{1}{2}}. \tag{3.18}$$

and δ, Re and We are the wave, Reynolds and Weisssenberg numbers, respectively. Under the assumptions of long wavelength and low Reynolds number, equations. (3.13) and (3.14) after using equation (3.16) become

$$\frac{\partial p}{\partial x} = \frac{\partial}{\partial y}\left[1 + \frac{(n-1)}{2}We^2\left(\frac{\partial^2 \psi}{\partial y^2}\right)^2\right]\frac{\partial^2 \psi}{\partial y^2}, \tag{3.19}$$

$$\frac{\partial p}{\partial y} = 0. \tag{3.20}$$

Eliminating pressure p from equati0ns (3.19) and (3.20) we finally get

$$\frac{\partial^2}{\partial y^2}\left[1+\frac{(n-1)}{2}We^2\left(\frac{\partial^2\psi}{\partial y^2}\right)^2\right]\frac{\partial^2\psi}{\partial y^2}=0 .$$ (3.21)

3.4 Rate of volume flow and boundary conditions

In laboratory frame, the dimensional volume flow rate is

$$Q = \int_{\bar{h}_2(\bar{X},\bar{t})}^{\bar{h}_1(\bar{X},t)} \bar{U}\left(\bar{X},\bar{Y},\bar{t}\right)d\bar{y} .$$ (3.22)

in which $\bar{h}_1$ and $\bar{h}_2$ are functions of $\bar{X}$ and $\bar{t}$, the above expression in wave frame becomes

$$q = \int_{\bar{h}_2(\bar{x})}^{\bar{h}_1(\bar{x})} \bar{u}\left(\bar{x},\bar{y}\right)d\bar{y},$$ (3.23)

where $\bar{h}_1$ and $\bar{h}_2$ are functions of $\bar{x}$ alone. Form equation (3.7), (3.22) and (3.23) we can write

$$Q = q + c\bar{h}_1\left(\bar{x}\right) - c\bar{h}_2\left(\bar{x}\right).$$ (3.24)

the Time- averaged flow over a period T at a fixed position $\bar{X}$ is

$$\theta = \frac{1}{T}\int_0^T Qdt,$$ (3.25)

Upon making use of equation (3.24)into equation(3.25) and then integrating one can write

$$\theta = F + c\bar{d}_1 + c\bar{d}_2 \ ,$$ (3.26)

If we define the dimensionless mean flows $\bar{Q}$ in the laboratory frame and q in the wave frame, according to

$$\bar{Q} = \frac{\theta}{c\bar{d}_1}, \quad q = \frac{F}{c\bar{d}_1},$$ (3.27)

one finds that equation. (3.26) becomes

$$\bar{Q} = q + 1 + d .$$ (3.28)

in which

$$F = \int_{h_2(x)}^{h1(x)} \frac{\partial \psi}{\partial y} dy = \psi\left(h_1(x)\right) - \psi\left(h_2(x)\right). \tag{3.29}$$

We note that $h_1(x)$ and $h_2(x)$ represent the dimensionless form of the surfaces of the peristaltic walls

$$h_1(x) = 1 + a\cos 2\pi x, \tag{3.30}$$

$$h_2(x) = -d - b\cos\left(2\pi x + \phi\right). \tag{3.31}$$

In wave frame the boundary conditions in terms of streams function ψ are

$$\psi = \frac{q}{2} at \ y = h_1(x), \tag{3.32}$$

$$\psi = \frac{-q}{2} at \ y = h_2(x), \tag{3.33}$$

$$\frac{\partial \psi}{\partial y} + \beta \frac{\partial^2 \psi}{\partial y^2} = -1 \ at \ y = h_1(x), \tag{3.34}$$

$$\frac{\partial \psi}{\partial y} - \beta \frac{\partial^2 \psi}{\partial y^2} = -1 \ at \ y = h_2(x). \tag{3.35}$$

where q is the flux, $\beta = \dfrac{\sqrt{k}}{\alpha d_1} = \dfrac{\sqrt{Da}}{\alpha}$ is permeability parameter including slip, Da is the Darcy number, α is the slip parameter and k is the permeability parameter α.

The boundary conditions (3.34) and (3.35) are introduced at the permeable walls of the channel following Saffman (1971). When the flow takes place past a permeable bed, it is well known that the usual no-slip condition is not valid at the nominal surface of the porous bed (Beavers and Joseph, 1967). The slip at the permeable wall is presented through a slip condition formulated by Saffman which is an improved condition to the Beavers and Joseph slip condition.

3.5 Perturbation solution

For perturbation solution, we expand ψ , q and p as

$$\psi = \psi_0 + We\psi_1 + O\left(We^2\right), \tag{3.36}$$

$$q = q_0 + Weq_1 + O\left(We^2\right), \tag{3.37}$$

$$p = p_0 + We p_1 + O(We^2).$$ (3.38)

3.5.1 System of order We^0

$$\frac{\partial^4 \psi_0}{\partial y^4} = 0,$$ (3.39)

$$\frac{\partial p_0}{\partial x} = \frac{\partial^3 \psi_0}{\partial y^3},$$ (3.40)

$$\psi_0 = \frac{q_0}{2} \text{ at } y = h_1(x),$$ (3.41)

$$\psi_0 = \frac{-q_0}{2} \text{ at } y = h_2(x),$$ (3.42)

$$\frac{\partial \psi_0}{\partial y} + \beta \frac{\partial^2 \psi_0}{\partial y^2} = -1 \text{ at } y = h_1(x),$$ (3.43)

$$\frac{\partial \psi_0}{\partial y} - \beta \frac{\partial^2 \psi_0}{\partial y^2} = -1 \text{ at } y = h_2(x).$$ (3.44)

3.5.2 System of order We^1

$$\frac{\partial p_1}{\partial y^4} + \frac{(n-1)}{2} \frac{\partial^2}{\partial y^2} \left[\left(\frac{\partial^2 \psi_0}{\partial y^2} \right)^3 \right] = 0,$$ (3.45)

$$\frac{\partial p_1}{\partial y} = \frac{\partial^3 \psi_1}{\partial y^3} + \frac{(n-1)}{2} \frac{\partial}{\partial y} \left[\left(\frac{\partial^2 \psi_0}{\partial y^2} \right)^3 \right]$$ (3.46)

$$\psi_1 = \frac{q_1}{2} \text{ at } y = h_1(x),$$ (3.47)

$$\psi_1 = \frac{-q_1}{2} \text{ at } y = h_2(x),$$ (3.48)

$$\frac{\partial \psi_1}{\partial y} + \beta \frac{\partial^2 \psi_1}{\partial y^2} = -1 \text{ at } y = h_1(x),$$ (3.49)

$$\frac{\partial \psi_1}{\partial y} - \beta \frac{\partial^2 \psi_1}{\partial y^2} = -1 \text{ at } y = h_2(x).$$ (3.50)

3.5.3 Solution for system of order We^0

Solving equation (3.39) and then using the boundary conditions (3.41) to (3.44) we have

$$\psi_0 = C_1 \frac{y^3}{3!} + C_2 \frac{y^2}{2!} + C_3 y + C_4. \tag{3.51}$$

where

$$C_1 = \frac{-12(q_0 + h_1 - h_2)}{(h_1 - h_2)^2 (h_1 - h_2 + 6\beta)},$$

$$C_2 = \frac{6(q_0 + h_1 - h_2)(h_1 + h_2)}{(h_1 - h_2)^2 [h_1 - h_2 + 6\beta]},$$

$$C_3 = -1 + \frac{6(q_0 + h_1 - h_2)[(h_1 - h_2) - h_1 h_2]}{(h_1 - h_2)^2 [h_1 - h_2 + 6\beta]},$$

$$C_4 = \left(\frac{q_0}{2} + h_1\right) + \frac{(q_0 + h_1 - h_2)[6\beta h_1(h_1 - h_2) - h_1^3 + 3h_1^2 h_2]}{(h_1 - h_2)^2 [h_1 - h_2 + 6\beta]}.$$

The expressions are the axial pressure gradient at this order is

$$\frac{\partial p_0}{\partial x} = \frac{-12(q_0 + h_1 - h_2)}{(h_1 - h_2)^2 (h_1 - h_2 + 6\beta)}. \tag{3.52}$$

Integrating equation (3.52) over per wavelength we get

$$\Delta P_{\lambda 0} = \int_0^1 \frac{dp_0}{dx} \, dx. \tag{3.53}$$

3.5.4 Solution for system of order We^1

Substituting the zeroth-order solution (3.51) into equation (3.45) and solving the resulting system along with the corresponding boundary conditions we find that

$$\psi_1 = F_1 \frac{y^3}{3!} + F_2 \frac{y^2}{2!} + F_3 y + F_4 + L_{11} y^5 + L_{12} y^4. \tag{3.54}$$

where

$$F_1 = \frac{-12[(q_1 + h_1 - h_2) - L_{17}(h_1 - h_2)]}{(h_1 - h_2)^2 [h_1 - h_2 + 6\beta]},$$

$$F_2 = \frac{6(h_1 + h_2)[q_1 + h_1 - h_2] + L_{18}}{(h_1 - h_2)^2 [h_1 - h_2 + 6\beta]},$$

$$F_3 = \frac{6(q_1 + h_1 - h_2)[\beta(h_1 - h_2) - h_1 h_2] + L_{19}}{(h_1 - h_2)^2 [h_1 - h_2 + 6\beta]},$$

$$F_4 = \frac{q_1}{2} - \frac{(q_1 + h_1 - h_2)[h_1^3 - 3h_1^2 h_1 + 6\beta h_1(h_1 - h_2)] + L_{21}}{(h_1 - h_2)^2 [h_1 - h_2 + 6\beta]}.$$

The axial pressure gradient and pressure drop per wavelength at respectively given by

$$\frac{dp_1}{dx} = -\frac{648(q_1 + h_1 - h_2)^3 (h_1 - h_2 - 2y)^2 + L_{22}}{(h_1 - h_2)^6 (h_1 - h_2 + 6\beta)^3}. \tag{3.55}$$

Integrating equation (3.55) over per wavelength we get

$$\Delta P_{\lambda 1} = \int_0^1 \frac{dp_1}{dx} dx. \tag{3.56}$$

Perturbation series solution up to second order for stream function ψ, velocity u, pressure gradient $\frac{dp}{dx}$ and pressure rise ΔP_λ may be summarized as

$$\psi = C_1 \frac{y^3}{3!} + C_2 \frac{y^2}{2!} + C_3 y + C_4 + We\left(F_1 \frac{y^3}{3!} + F_2 \frac{y^2}{2!} + F_3 y + F_4 + L_{11} y^5 + L_{12} y^4 \right), \tag{3.57}$$

$$u = C_1 \frac{y^2}{2!} + C_2 y + C_3 + We\left(F_1 \frac{y^2}{2!} + F_2 y + F_3 + 5L_{11} y^4 + 4L_{12} y^3 \right), \tag{3.58}$$

$$\frac{dp}{dx} = \frac{dp_0}{dx} + We\frac{dp_1}{dx}, \tag{3.59}$$

$$\Delta P_\lambda = \Delta P_{\lambda 0} + We\Delta P_{\lambda 1}. \tag{3.60}$$

The frictional force, at $y = h_1$ and $y = h_2$ denoted by

$$F_{\lambda 1} = \int_0^1 -h_1^2 \left(\frac{dp}{dx} \right) dx, \tag{3.61}$$

$$F_{\lambda 2} = \int_0^1 -h_2^2 \left(\frac{dp}{dx} \right) dx. \tag{3.62}$$

3.6 Expressions for wave shape:

The non-dimensional expressions for the five considered wave forms are given by the following equations:

1. Sinusoidal wave:

$$h(x) = 1 + a\sin(x),$$ (3.63)

2. Triangular wave:

$$h(x) = 1 + a\left\{\frac{8}{\pi^3}\sum_{m=1}^{\infty}\frac{(-1)^{m+1}}{(2m-1)^2}\sin[(2m-1)x]\right\},$$ (3.64)

3. Square wave:

$$h(x) = 1 + a\left\{\frac{4}{\pi}\sum_{m=1}^{\infty}\frac{(-1)^{m+1}}{(2m-1)}\cos[(2m-1)x]\right\},$$ (3.65)

4. Saw tooth wave:

$$h(x) = 1 + a\left\{\frac{8}{\pi^3}\sum_{m=1}^{\infty}\frac{\sin[2m\pi x]}{m}\right\},$$ (3.66)

5. Trapezoidal wave:

$$h(x) = 1 + a\left\{\frac{32}{\pi^2}\sum_{m=1}^{\infty}\frac{\sin\frac{\pi}{8}(2m-1)}{(2m-1)^2}\sin[(2m-1)x]\right\}.$$ (3.67)

Appendix

$$L_{11} = \frac{-(n-1)c_1^3}{40},$$

$$L_{12} = \frac{-(n-1)c_1^2 c_2}{8},$$

$$L_{13} = 5L_{11}\left(h_1^4 + 4\beta h_1^3\right) + 4L_{12}\left(h_1^3 + 3\beta h_1^2\right),$$

$$L_{14} = 5L_{11}\left(h_2^4 - 4\beta h_2^3\right) + 4L_{12}\left(h_2^3 - 3\beta h_2^2\right),$$

$$L_{15} = \frac{L_{14} - L_{13}}{\left(h_1 - h_2 + 2\beta\right)},$$

$$L_{16} = L_{11}\left(h_1^4 + h_1^3 h_2 + h_1^2 h_2^2 + h_1^2 h_2^2 + h_1 h_2^3 + h_2^4\right) + L_{12}\left(h_1 + h_2\right)\left(h_1^2 + h_2^2\right) - L_{13},$$

$$L_{17} = \frac{1}{2}\left[L_{15}(h_2 - h_1 - 2\beta)\right] + L_{16},$$

$$L_{18} = L_{15}\left(h_1 - h_2\right)^2\left[h_1 - h_2 + 6\beta\right] - 6L_{17}\left(h_1^2 - h_2^2\right),$$

$$L_{19} = \left(h_1 - h_2\right)^2\left(h_1 - h_2 + 6\beta\right)\left[-1 - L_{13} - L_{15}\left(h_1 + \beta\right)\right] - 6L_{17}\left(h_1 - h_2\right)\left[\beta(h_1 - h_2) - h_1 h_2\right],$$

$$L_{20} = h_1\left[1 + L_{13} + L_{15}(h_1 + \beta)\right] - L_{15}\frac{h_1^2}{2} - L_{12}h_1^4 - L_{11}h_1^5,$$

$$L_{21} = L_{20}\left(h_1 - h_2\right)^2(h_1 - h_2 + 6\beta) + L_{17}(h_1 - h_2)\left[h_1^3 - 3h_1^2 h_2 + 6h_1\beta(h_1 - h_2)\right],$$

$$L_{22} = (h_1 - h_2)^4(h_1 - h_2 + 6\beta)^2\left[12(L_{17}(h_1 - h_2) - (q_1 + h_1 - h_2))\right].$$

3.7 Results and discussion:

The variation of velocity u with y is calculated from equation (3.58) different values of the average volume rate $\overline{Q}$ with x=1, a=0.5, b=0.5, d=1, $\phi = \frac{\pi}{6}$, β =0.2, n=0.1 and We =0.01 and is depicted in figure (3.2). It is observed that velocity u increases with increasing the average volume flow rate $\overline{Q}$.

The variation between velocity u with y is drawn in figure (3.3) for different values of the phase difference ϕ with x=1, a=0.6, b=0.8, d=1, $\overline{Q}$=3, β =0.3, n=0.1 and We =0.01. It is concluded that the velocity decreases by increasing phase difference ϕ in the lower half of the channel and in the upper half of the channel.

The relationship between velocity u with y is shown for different values of the permeability parameter including slip β with x=1, a=0.6, b=0.8, d=1, $\phi = \frac{\pi}{6}$, $\overline{Q}$ =3, n=0.1 and We =0.01 in figure (3.4). It is observed that the velocity u increases with increasing permeability parameter including slip β near the walls. However, u decreases by increasing β near the centre of the channel $\frac{dp}{dx}$.

In figure (3.5) the relation between velocity u with y is plotted for different values of Weissenberg number We with x=1, a=0.5, b=0.6, d=1, $\phi = \dfrac{\pi}{3}$, $\overline{Q}$=2, n=0.2 and β =0.2. It is observed that the velocity increases with increasing Weissenberg number We .

The variation of pressure gradient $\dfrac{dp}{dx}$ with x is calculated from equation (3.59) for different values of volume flow rate $\overline{Q}$ with a=0.5, b=0.5, n=0.5, d=1, $\phi = \dfrac{\pi}{6}$, β =0.2 and We =0.01 and is plotted in figure (3.6). It is observed that pressure gradient $\dfrac{dp}{dx}$ decreases with increasing the volume flow rate $\overline{Q}$.

The variation of pressure gradient $\dfrac{dp}{dx}$ with x for different values of phase difference ϕ with a=0.5, b=0.5, d=1, $\overline{Q}$=0.7, β =0.2, n=0.1 and We =0.01 and is plotted in figure (3.7). It is observed that pressure gradient $\dfrac{dp}{dx}$ decreases with the increase in the phase difference and point of maximum decreases with increasing ϕ.

The relation between pressure gradient $\dfrac{dp}{dx}$ and x is drawn in figure (3.8) for different values of Weissenberg number We with a=0.5, b=0.5, d=1, $\overline{Q}$=0.1, n=0.1, $\phi = \dfrac{\pi}{6}$ and β=0.1.It is found that the pressure gradient $\dfrac{dp}{dx}$ decreases with increasing Weissenberg number We .

The variation between pressure gradient $\dfrac{dp}{dx}$ with x is drawn in figure (3.9) for different values of permeability parameter including slip β with a=0.5, b=0.5, d=0.5, $\overline{Q}$=0.7, n=0.5, $\phi = \dfrac{\pi}{6}$ and We =0.01. It is concluded that the pressure gradient $\dfrac{dp}{dx}$ increases with decreasing permeability parameter including slip β.

The relation between pressure gradient $\dfrac{dp}{dx}$ and x is plotted in figure (3.10)

for different values of Weissenberg number We with a=0.6, b=0.6, d=0.6, $\overline{Q}$=1,

n=0.1, $\phi = \dfrac{\pi}{6}$ and β=0.1. It is observed that the pressure gradient $\dfrac{dp}{dx}$ decreases with

increasing Weissenberg number We .

From figure (3.11) the variation pressure gradient $\dfrac{dp}{dx}$ with x is depicted for

different channel width d with a=0.8, b=0.5, $\overline{Q}$=1, $\phi = \dfrac{\pi}{6}$, β=0.1, n=0.1 and We

=0.01. It is concluded that the pressure rise decreases by increasing channel width d.

The relation between pressure gradient $\dfrac{dp}{dx}$ and x is shown (3.12) for various

values of wave amplitude a with b=0.6, d=0.5, $\overline{Q}$=1, $\phi = \dfrac{\pi}{6}$, β=0.1, n=1and We

=0.01. It is observed that the pressure gradient $\dfrac{dp}{dx}$ increases with increasing wave

amplitude a.

The variation on pressure rise ΔP_λ with mean flux $\overline{Q}$ is calculated from

equation (3.60) and is drawn figure (3.13) for different values of permeability

parameter β with a=0.5, b=0.5, d=1, $\phi = \dfrac{\pi}{6}$, n=0.1and We =0.01. We conclude that

for values of $\overline{Q}$ between 0.6 and 0.7, the pumping curves intersect at a point (0.67, -

0.25) and this behaviour occurs for co-pumping range ΔP_λ<0. For ΔP_λ>0 and for

given mean flux $\overline{Q}$, the pressure rise ΔP_λ decreases with increasing permeability

parameter β.

The relation between pressure rise ΔP_λ with mean flux $\overline{Q}$ is plotted in figure

(3.14) for different values of phase difference ϕ with a=0.8, b=0.4, d=1, β =0.01,

n=0.4 and We =0.01. We found that for values of $\overline{Q}$ between 1.2 and 1.4, the

pumping curves intersect at a point (1.25, -2) and this behaviour occurs for co-pumping range $\Delta P_\lambda < 0$. For $\Delta P_\lambda > 0$ and for a given mean flux $\overline{Q}$, the pressure rise ΔP_λs decreases with increasing phase differenceϕ. For a given ΔP_λ the flux decreases with increasing ϕ.

Figure (3.15) shows the relation between pressure rise ΔP_λ with mean flux $\overline{Q}$ for various values of Weissenberg number We with a=0.5, b=0.5, d=0.5, $\phi = \frac{\pi}{6}$, n=0.4 and β =0.1. We found that for values of $\overline{Q}$ between 0.6 and 0.8, the pumping curves intersect at a point (0.7, 1). For a given mean flux $\overline{Q}$, the pressure rise ΔP_λ increases with decreasing Weissenberg number We above this point and opposite behaviour is observed below this point.

The relationship between pressure rise ΔP_λ with mean flux $\overline{Q}$ is drawn figure (3.16) for different values n with a=0.5, b=0.5, d=0.5, ϕ=π/6, β =0.1 and We =0.03. We conclude that the pumping curves intersect at a point (0.9, 0). For a given flux $\overline{Q}$, the pressure rise ΔP_λ increases with increasing n below this point and opposite behaviour is observed above this point.

The relationship between pressure rise ΔP_λ with mean flux $\overline{Q}$ is drawn figure (3.17) for different values of channel width d with a=0.5, b=0.5, ϕ=π/6, β =0.1, n=0.1 and We =0.03. We conclude that for values of $\overline{Q}$ between 0.6 and 0.8, the pumping curves intersect at a point (0.7, -1). For a given flux $\overline{Q}$, the pressure rise ΔP_λ increases with increasing channel width d below this point and opposite behaviour is observed above this point.

The variation of pressure rise ΔP_λ with mean flux $\overline{Q}$ is depicted figure (3.18) for different values of wave amplitude a with b=0.5, d=1, $\phi = \frac{\pi}{6}$, β =0.1, n=0.1 and We =0.01. We infer that for values of $\overline{Q}$ between 1.2 and 1.4, the pumping curves intersect at a point (1.3, -2.2). For a given flux $\overline{Q}$, the pressure rise ΔP_λ decreases

113

with increasing wave amplitude a below this point and opposite behaviour is observed above this point.

The relation between pressure rise ΔP_λ with mean flux $\bar{Q}$ is depicted figure (3.19) for different values of wave amplitude b with a=0.6, d=1, $\phi=\pi/6$, $\beta=0.01$, n=0.1 and $We=0.01$. We found that for values of $\bar{Q}$ between 1.4 and 1.6, the pumping curves intersect at a point (1.39, -2.2). For a given flux $\bar{Q}$, the pressure rise ΔP_λ decreases with increasing wave amplitude b below this point and opposite behaviour is concluded above this point.

The variation of frictional forces $F_{\lambda 1}$ and $F_{\lambda 2}$ with $\bar{Q}$ is shown figures (3.20) to (3.30). We observe that the frictional force shows opposite behaviour to that of pressure rise for the corresponding variations in the physical parameters ϕ, d, β and We.

3.7.1 Trapping phenomena

Another interesting phenomenon in the peristaltic motion is trapping. In the wave frame streamlines under certain conditions split to trap a bolus which moves as a whole with the speed of the wave. Streamlines are plotted to observe of effects of various parameters on trapping both the Carreau and Newtonian fluids.

The stream lines are drawn in figure (3.31) with different values of volume flow rate $\bar{Q}$ at a=0.5, b=0.5, d=1, $\beta=0.1$, $\phi=0$ and $We=0.01$.. The size and shape of the trapped bolus is independent of Carreau fluid parameters. The trapping is found to be dependent on the values of volume flow rate $\bar{Q}$ both in Carreau and Newtonian fluids, the size of trapping bolus increases with increasing volume flow rate and trapping bolus disappears for $\bar{Q}=1$.

The stream lines are plotted in figure (3.32) for various values of the phase difference ϕ with a=0.5, b=0.5, d=1, $\beta=0.1$, $\bar{Q}=1.8$ and $We=0.01$. The size and shape of the trapped bolus is independent of Carreau fluid parameters. The trapping is concluded to be dependent on the values of phase difference ϕ both in Carreau and

Newtonian fluids, the volume of the trapping bolus appearing in the central region for ϕ=0, moves towards left and decreases in volume as v increases.

The stream lines are drawn for various values of the permeability parameter including slip β and are shown in figure (3.33) with a=0.5, b=0.5, d=1, ϕ=0, $\overline{Q}$=1.8 and We =0.01. It is found that the volume of the trapping bolus increases upper half of the channel and decreases the lower half of the channel with increasing permeability parameter including slip β.

The stream lines are depicted for different values of the Weissenberg number We and are shown in figure (3.34) with a=0.5, b=0.5, d=1, ϕ=0, $\overline{Q}$=1.8 and β=0.2. It is observed that the volume of the trapping bolus increases lower half of the channel and decreases the upper half of the channel with increasing Weissenberg number We .

The stream lines are depicted in the wave frame for pumping for different values of channel width d in figure (3.35) with a=0.5, b=0.5, β=0.2, ϕ=0, $\overline{Q}$=1.8 and We =0.01. It is observed that the size of the trapping bolus decreases with increasing channel width d.

Stream lines are depicted in figures in (3.36) and (3.37) for different values of the average volume flow rate $\overline{Q}$ with a=0.5, b=0.5, d=1, n=0.5, ϕ=0,□ β=0.2 and We =0.01with the following wave forms a) sinusoidal wave, b) trapezoidal, c) triangular wave, d) square wave and e) sawthooth wave. We observed that the size of the bolus increases with increasing $\overline{Q}$ for the wave forms considered.

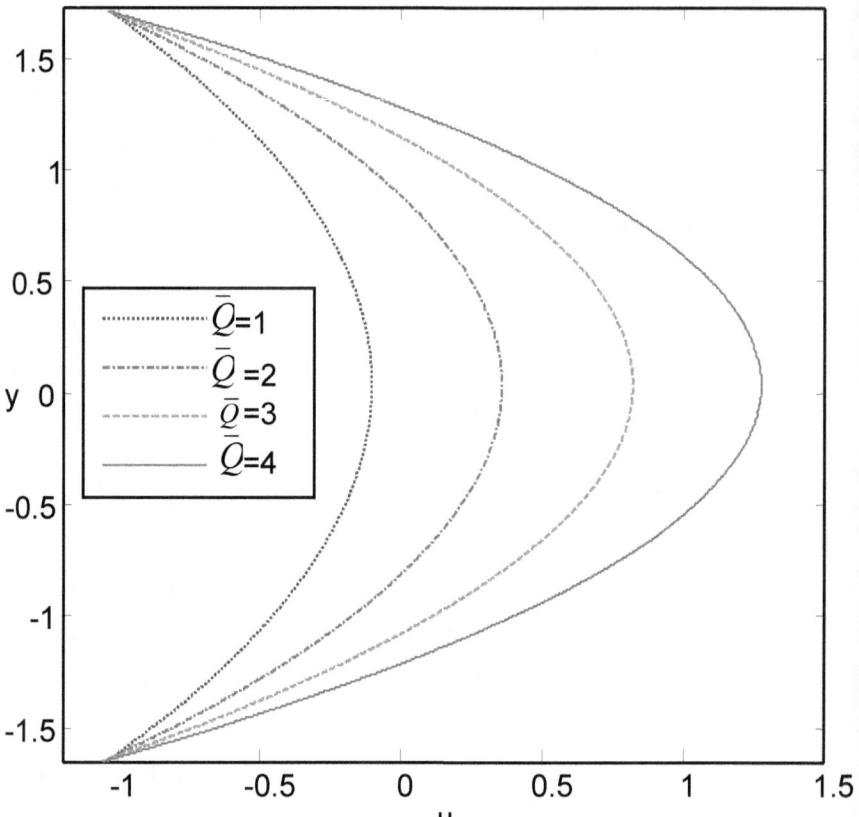

Figure 3.2: The velocity profiles with a=0.5, b=0.5, d=1, x=1, $\phi=\pi/6$, β=0.2, n=0.1 and We =0.01.

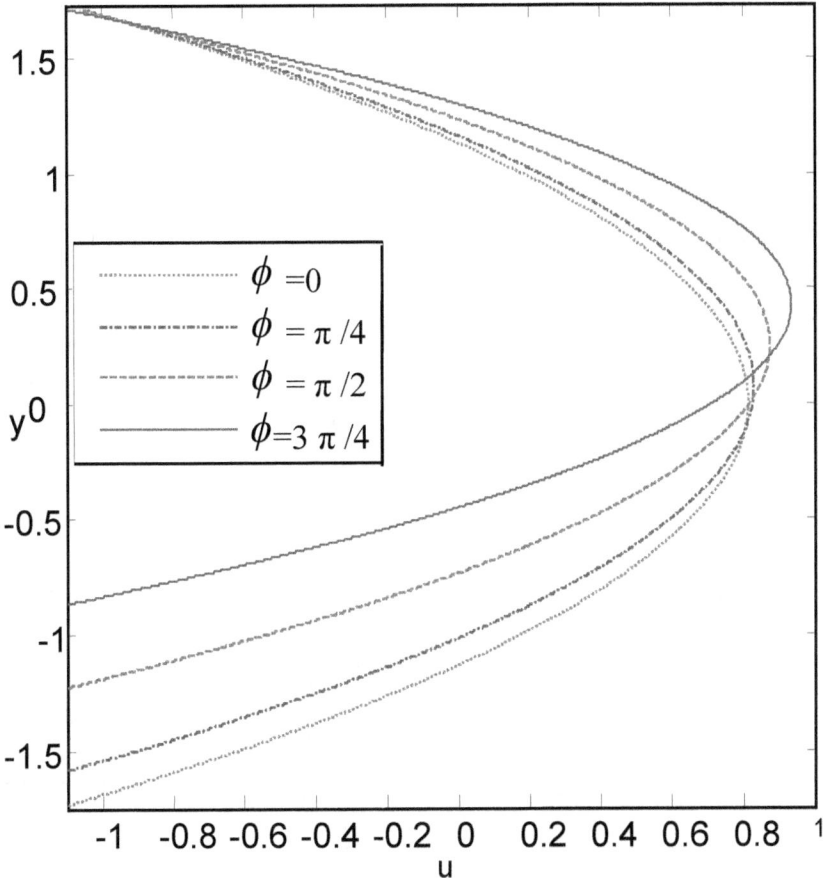

Figure 3.3: The velocity profiles with a=0.6, b=0.8, d=1, x=1, $\overline{Q}$=3, β=0.3, n=0.1 and We =0.01.

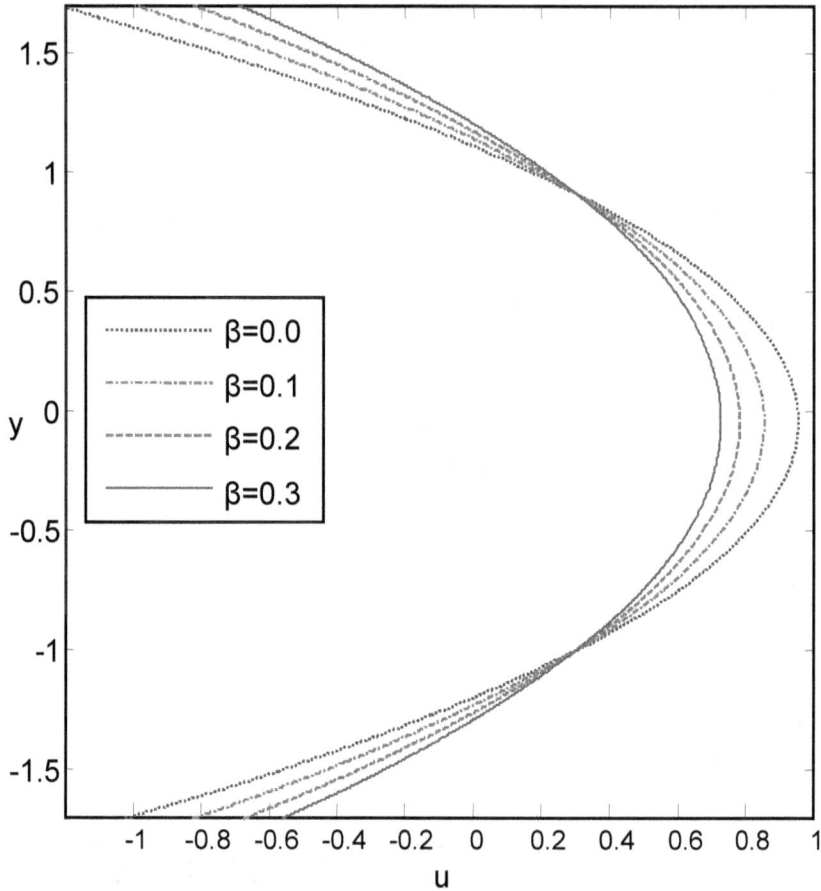

Figure 3.4: The velocity profiles with a=0.6, b=0.8, d=1, x=1, $\overline{Q}$=3, ϕ=π/6, n=0.1 and We =0.01.

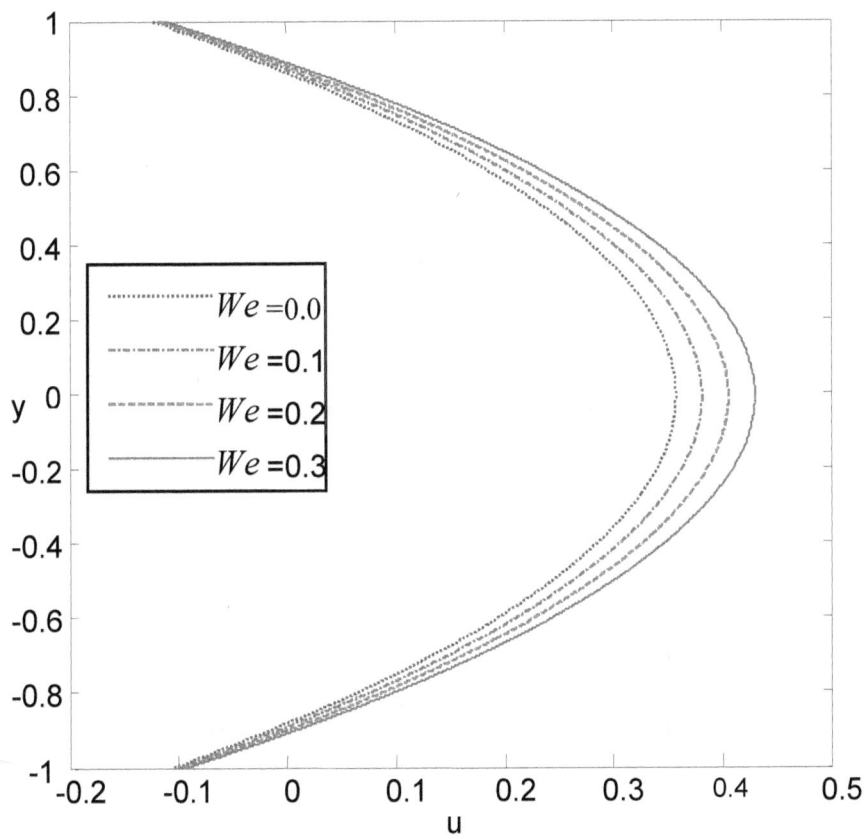

Figure 3.5: The velocity profiles with a=0.5, b=0.6, d=1, x=1, $\overline{Q}$=2, ϕ=π/6, β=0.2 and n=0.2.

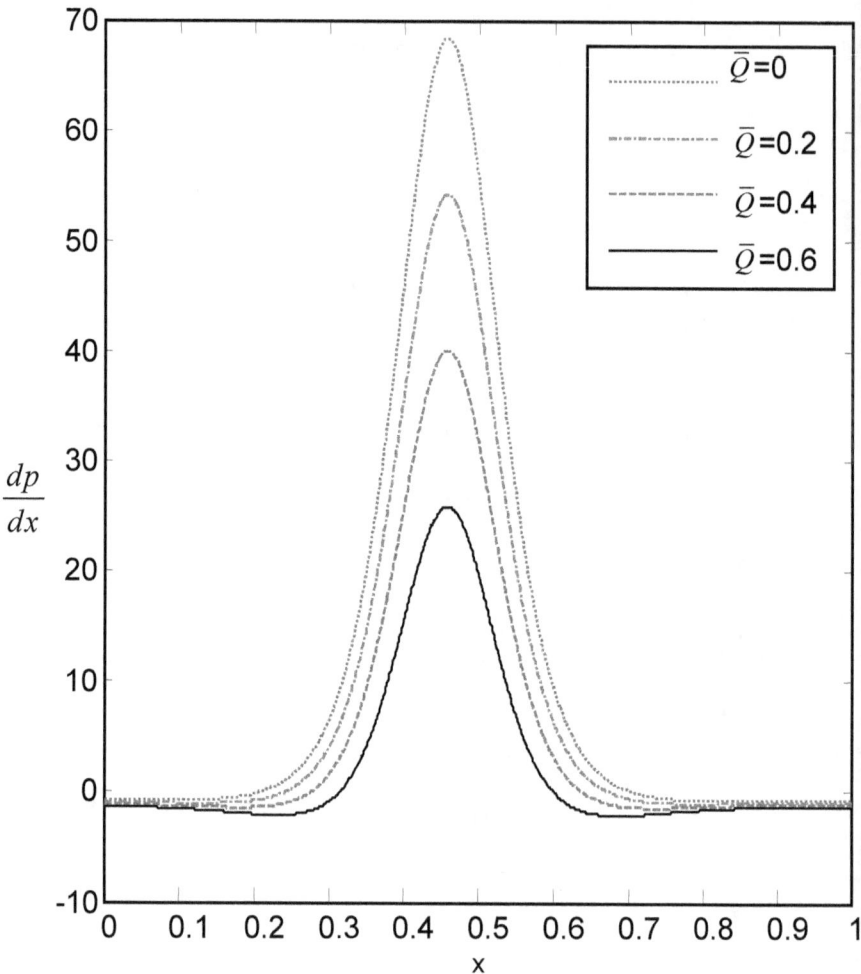

Figure 3.6: Various of pressure gradient $\dfrac{dp}{dx}$ with 'x' for different values of the average volume flow rate $\overline{Q}$ with a=0.5, b=0.5, d=1, $\phi = \pi / 6$, β=0.2 , n=0.5 and We =0.01.

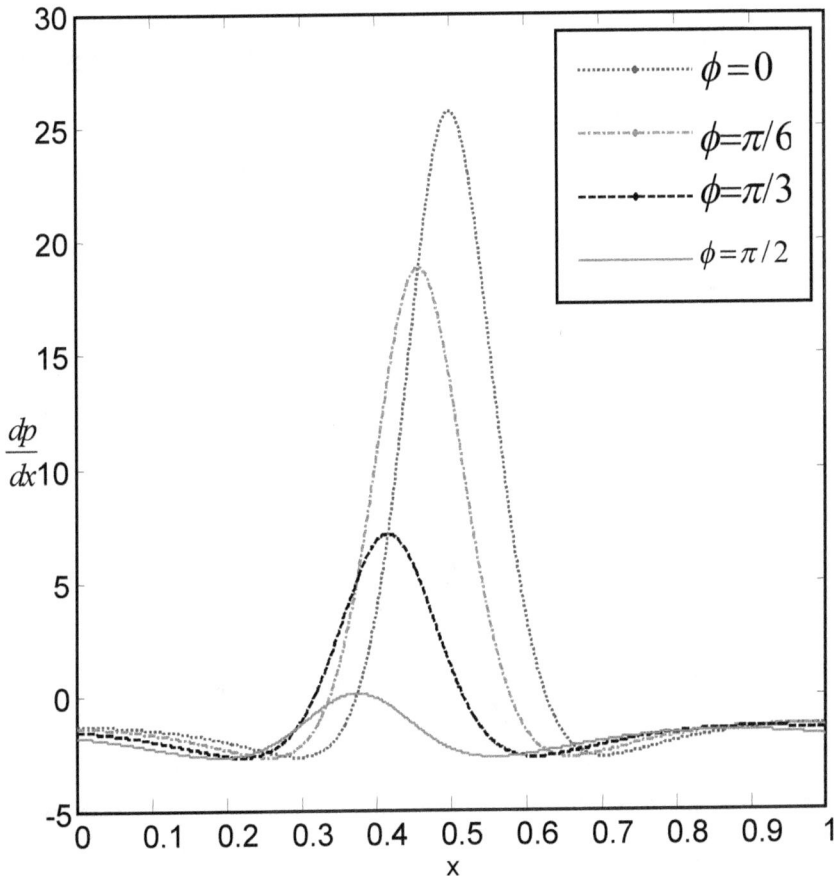

Figure 3.7: Various of pressure gradient $\dfrac{dp}{dx}$ with 'x' for different values of phase

difference ϕ with a=0.5, b=0.5, d=1, β=0.2, $\overline{Q}$=0.7, n=0.1 and

We =0.06.

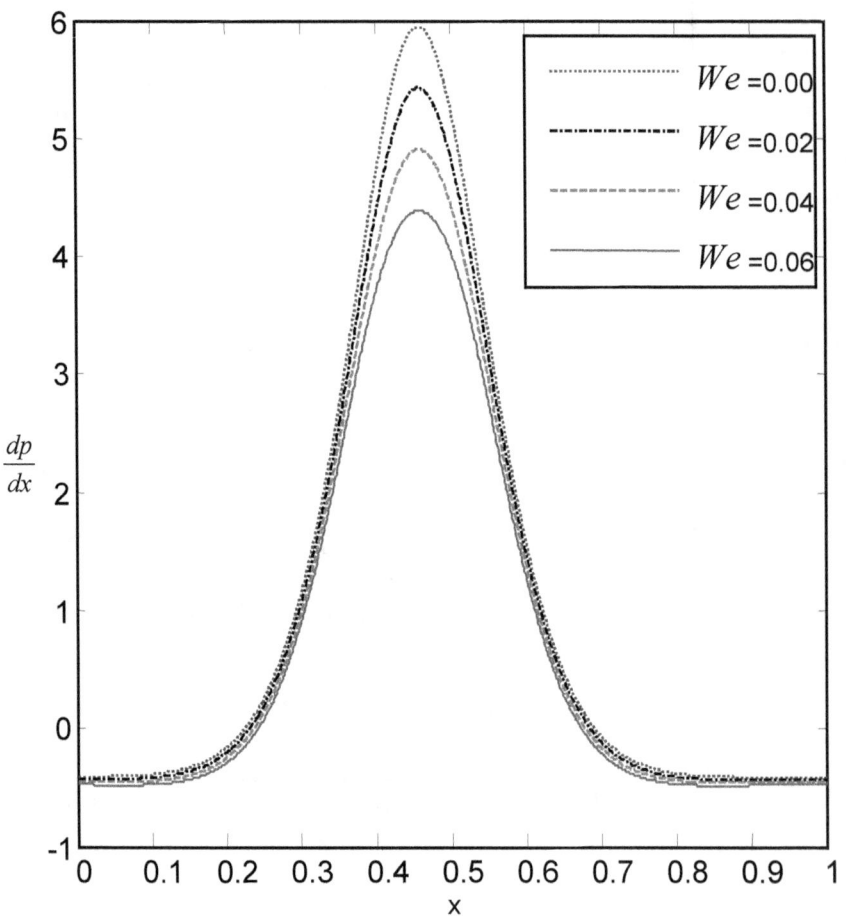

Figure 3.8: Various of pressure gradient $\dfrac{dp}{dx}$ with 'x' for different values of

Weissenberg number We with a=0.5, b=0.5, d=1, $\phi=\pi/6$, $\beta=0.1$, $\overline{Q}$ =0.1 and n=0.1.

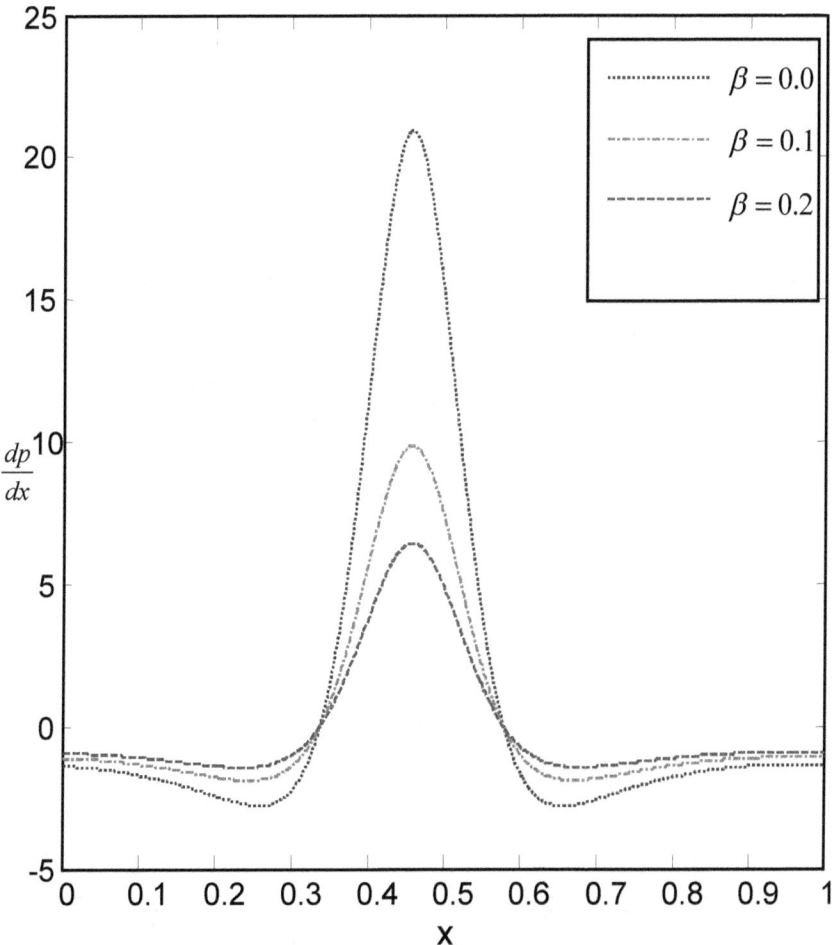

Figure 3.9: Various of pressure gradient $\dfrac{dp}{dx}$ with 'x' for different values of

permeability parameter β with   a=0.5, b=0.5, d=0.5, ϕ=π/6, $\overline{Q}$=0.7,

n=0.5 and We =0.01.

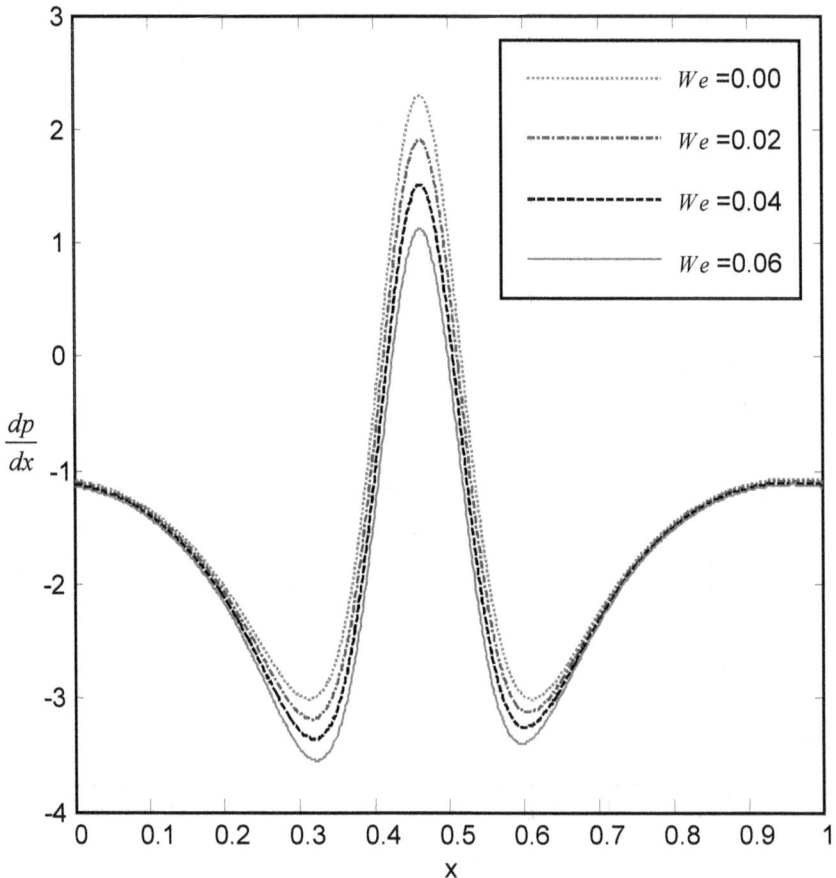

Figure 3.10: Various of pressure gradient $\frac{dp}{dx}$ with 'x' for different values of

Weissenberg number We with   a=0.6, b=0.6, d=0.6, ϕ=π/6, β =0.1, $\overline{Q}$
=1, and n=0.1.

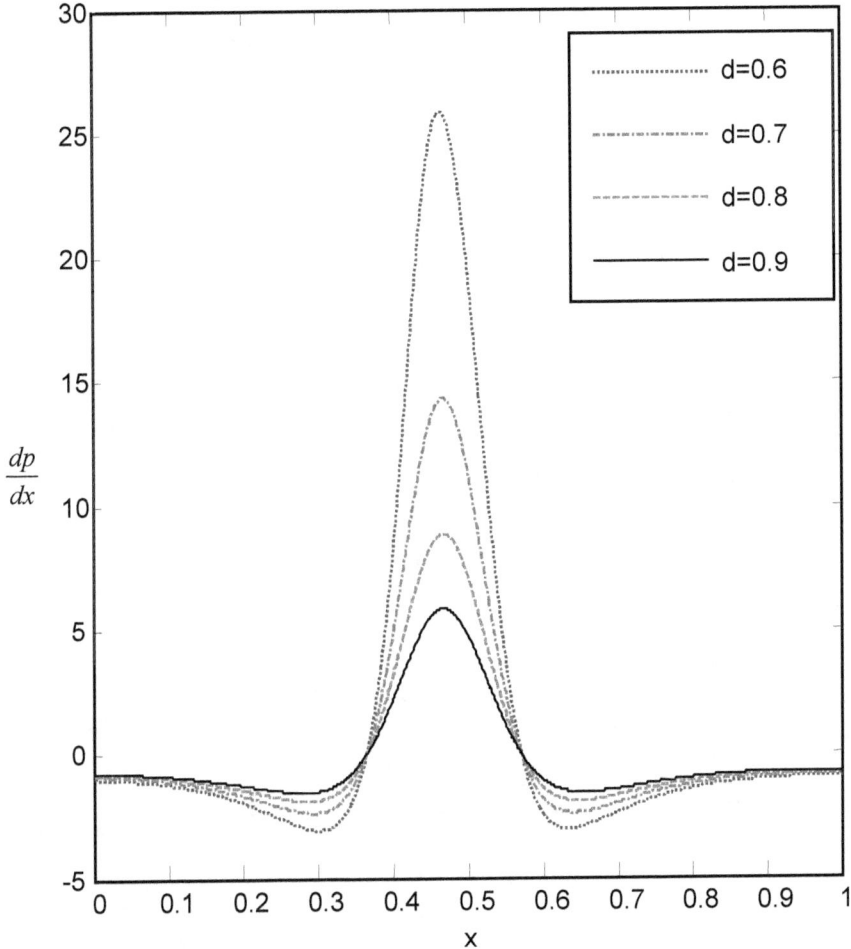

Figure 3.11: Various of pressure gradient $\dfrac{dp}{dx}$ with 'x' for different values of width

d with   a=0.8, b=0.5, ϕ=π/6, β =0.1, $\overline{Q}$=1, n=0.1 and We =0.01.

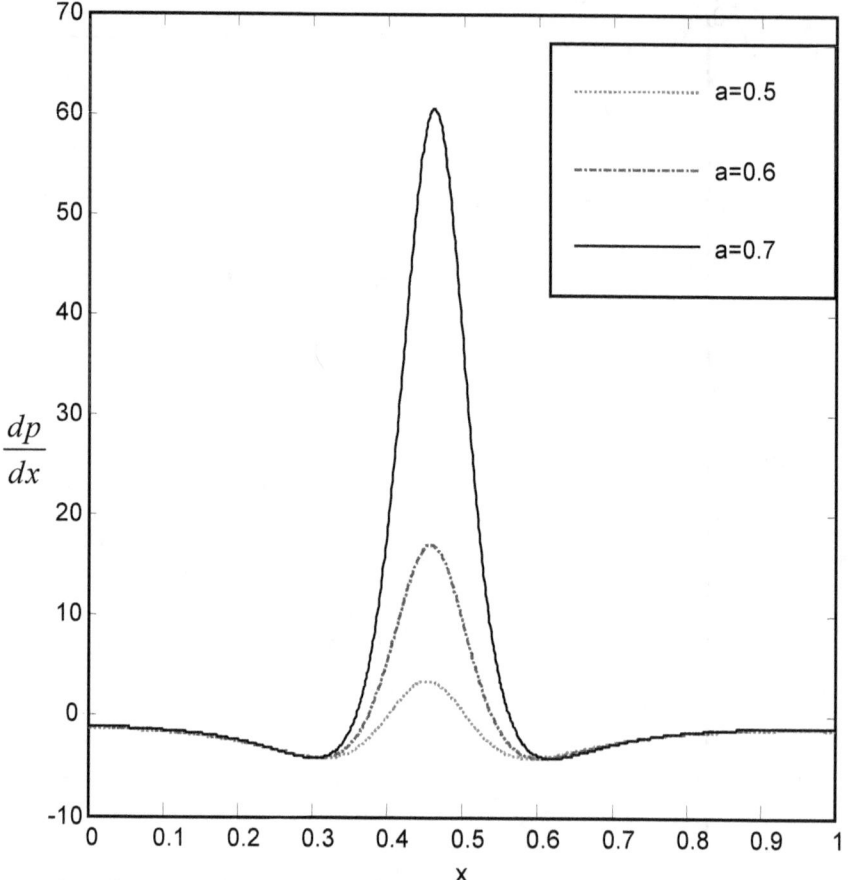

Figure 3.12: Various of pressure gradient $\dfrac{dp}{dx}$ with 'x' for different values of wave amplitude 'a' with b=0.6, d=0.5, ϕ=π/6, β =0.1, $\overline{Q}$=1, n=1 and We =0.01.

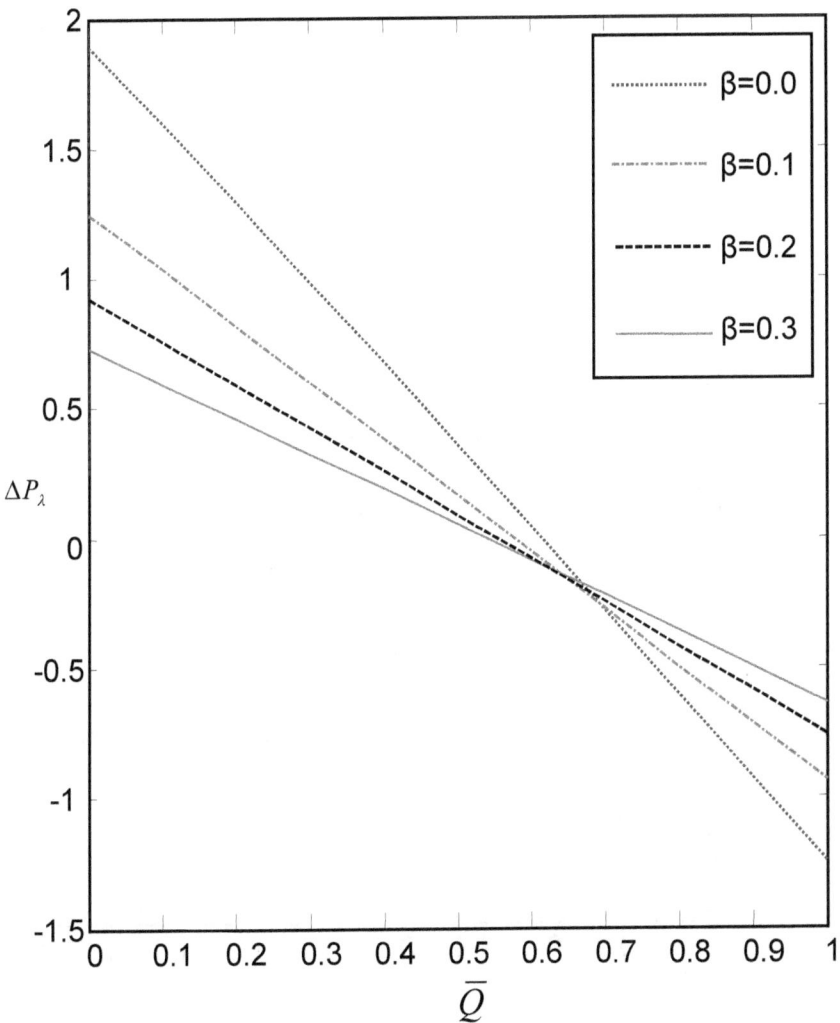

Figure 3.13: Variation of $\overline{Q}$ with ΔP_{λ} for different values of permeability parameter 'β' with a=0.5, b=0.5, d=1, $\phi=\pi/6$, n=0.1 and We =0.01.

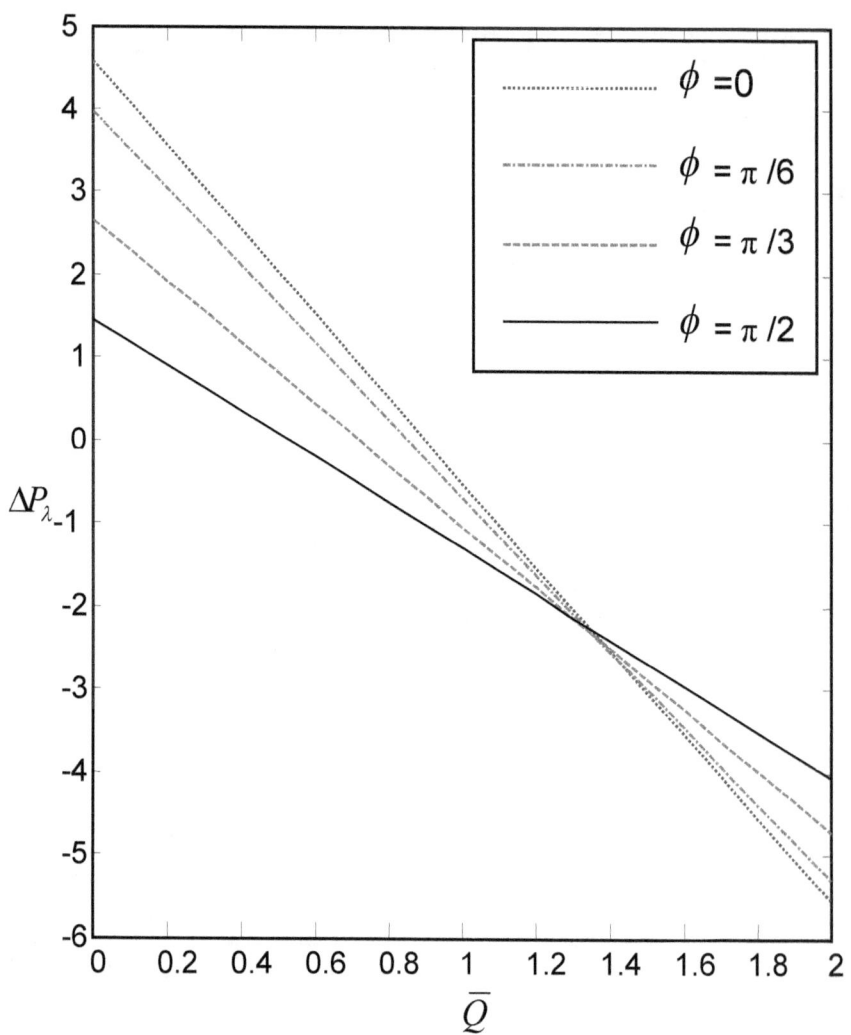

Figure 3.14: Variation of $\overline{Q}$ with ΔP_λ for different values of phase difference 'ϕ' with a=0.8, b=0.4, d=1, β=0.01, n=0.4 and We =0.01.

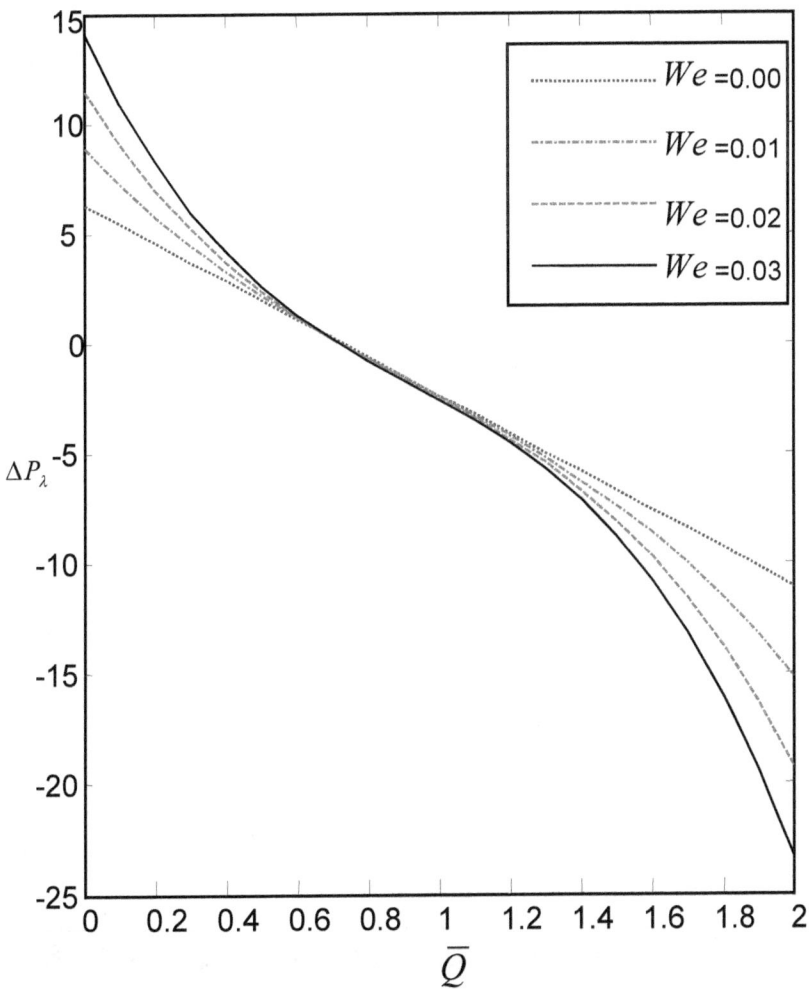

Figure 3.15: Variation of $\overline{Q}$ with ΔP_λ for different values of Weissenberg number 'We' with a=0.5, b=0.5, d=0.5, ϕ=π/6, β=0.1 and n=0.4.

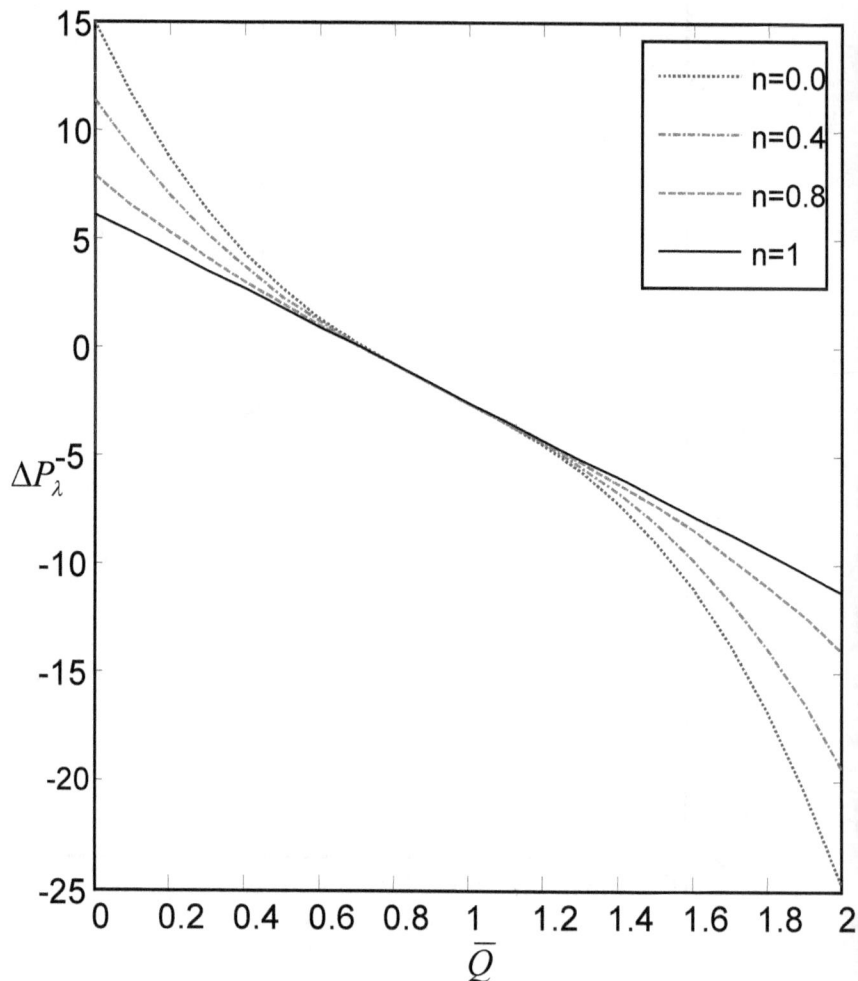

Figure 3.16: Variation of $\overline{Q}$ with ΔP_λ for different valuesn of 'n' with a=0.5, b=0.5, d=0.5, $\phi=\pi/6$, $\beta=0.1$ and We =0.03.

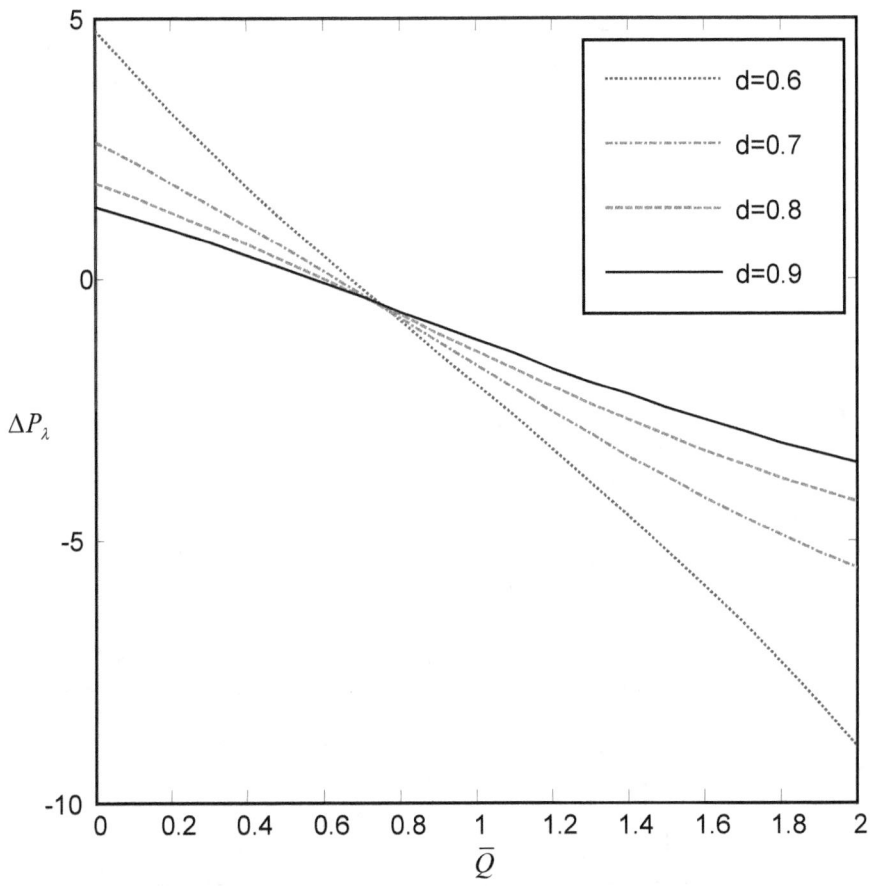

Figure 3.17: Variation of $\overline{Q}$ with ΔP_λ for different values of width 'd' with a=0.5, b=0.5, $\phi=\pi/6$, $\beta=0.1$, n=0.1 and We =0.03.

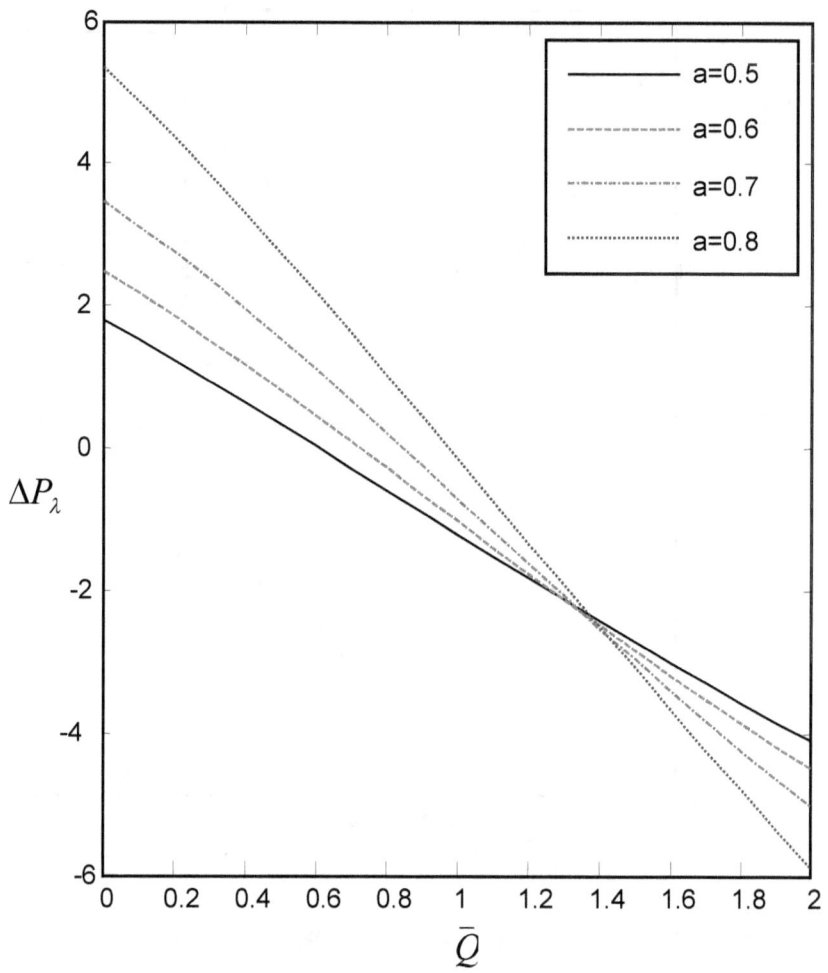

Figure 3.18: Variation of $\overline{Q}$ with ΔP_{λ} for different wave amplitude 'a' with b=0.5, d=1, $\phi=\pi/6$, $\beta=0.1$, n=0.1 and We =0.01.

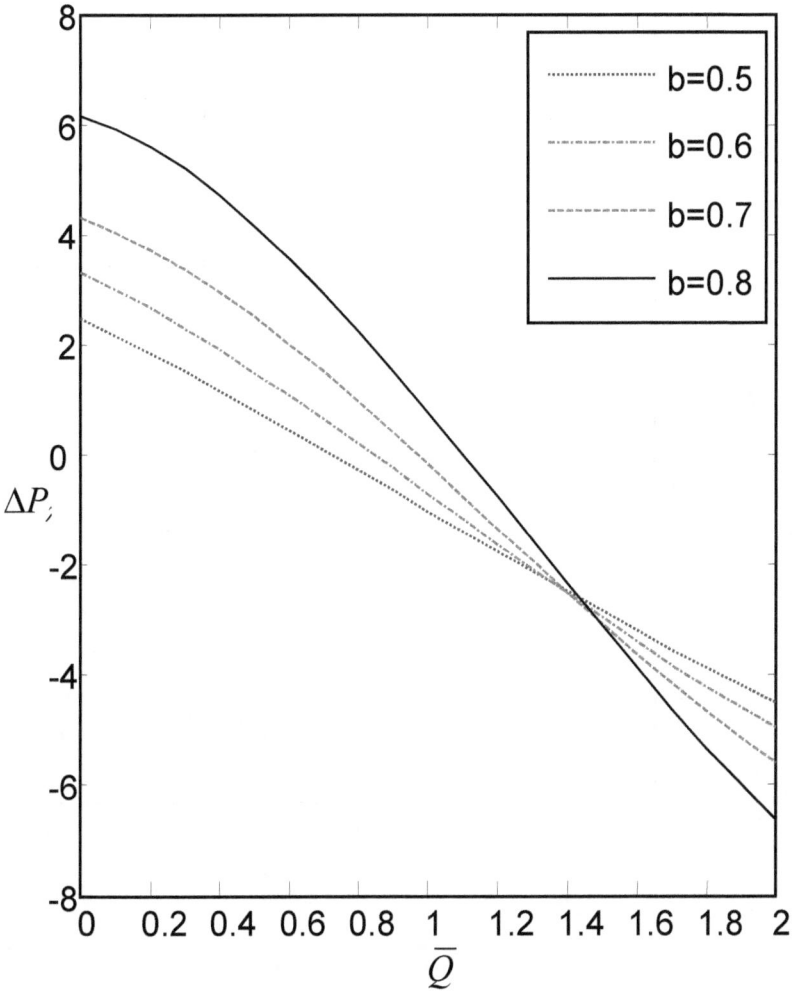

Figure 3.19: Variation of $\overline{Q}$ with ΔP_λ for different values of wave amplitude 'b' with a=0.6, d=1, $\phi=\pi/6$, $\beta=0.01$, n=0.1, and We =0.01.

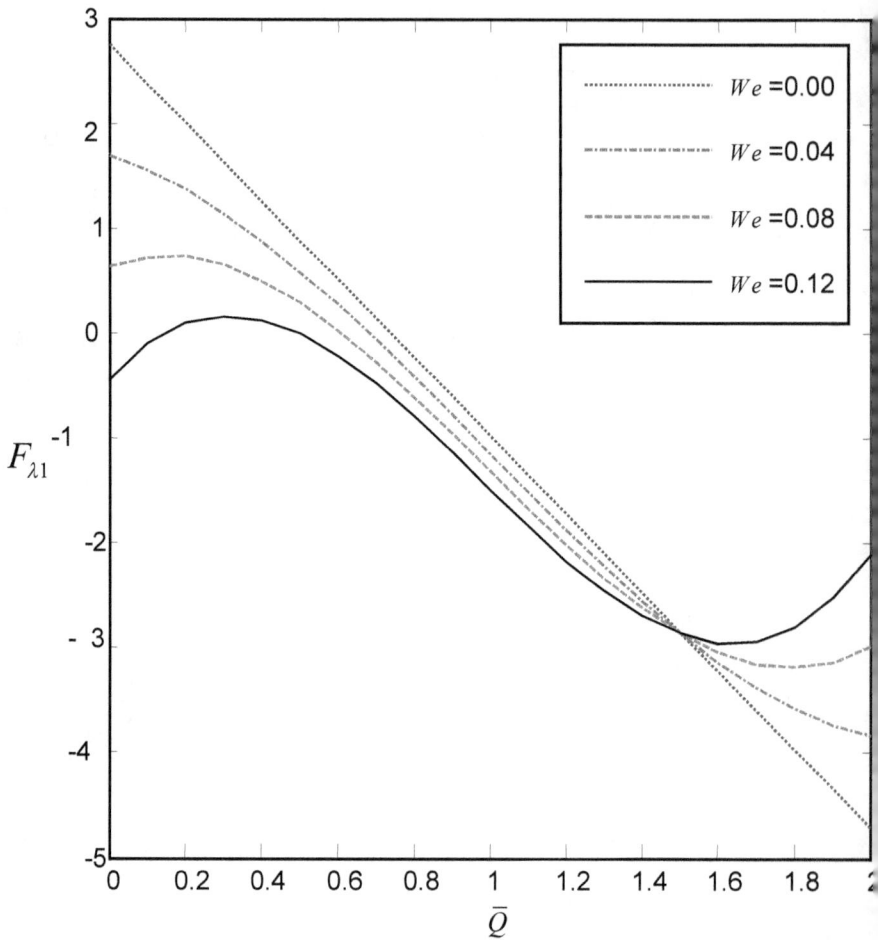

Figure 3.20: Various of frictional force $F_{\lambda 1}$ with $\bar{Q}$ at $y = h_1$ for different values of Weissenberg number 'We' with a=0.6, b=0.5, d=1, $\phi=\pi/6$, $\beta=0.01$ and n=0.1.

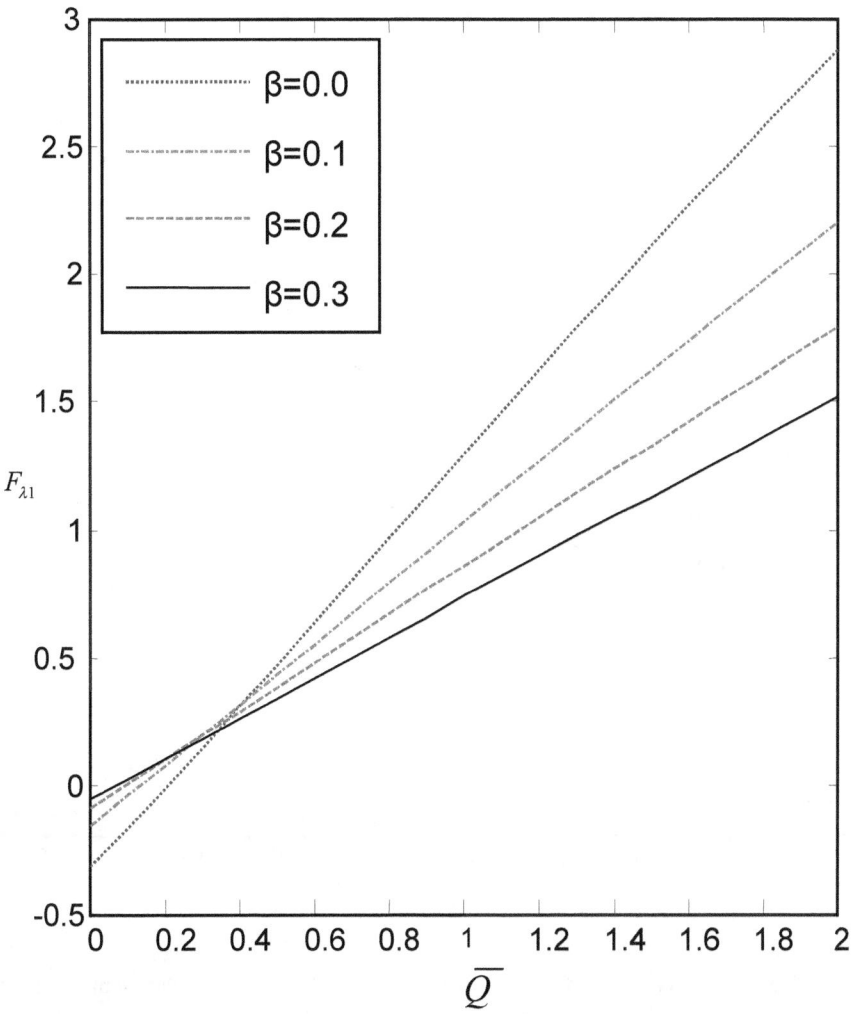

Figure 3.21: Various of frictional force $F_{\lambda 1}$ with $\overline{Q}$ at $y = h_1$ for different values of permeability parameter 'β' with a=0.6, b=0.5, d=1, $\phi=\pi/6$, n=0.1 and We =0.01.

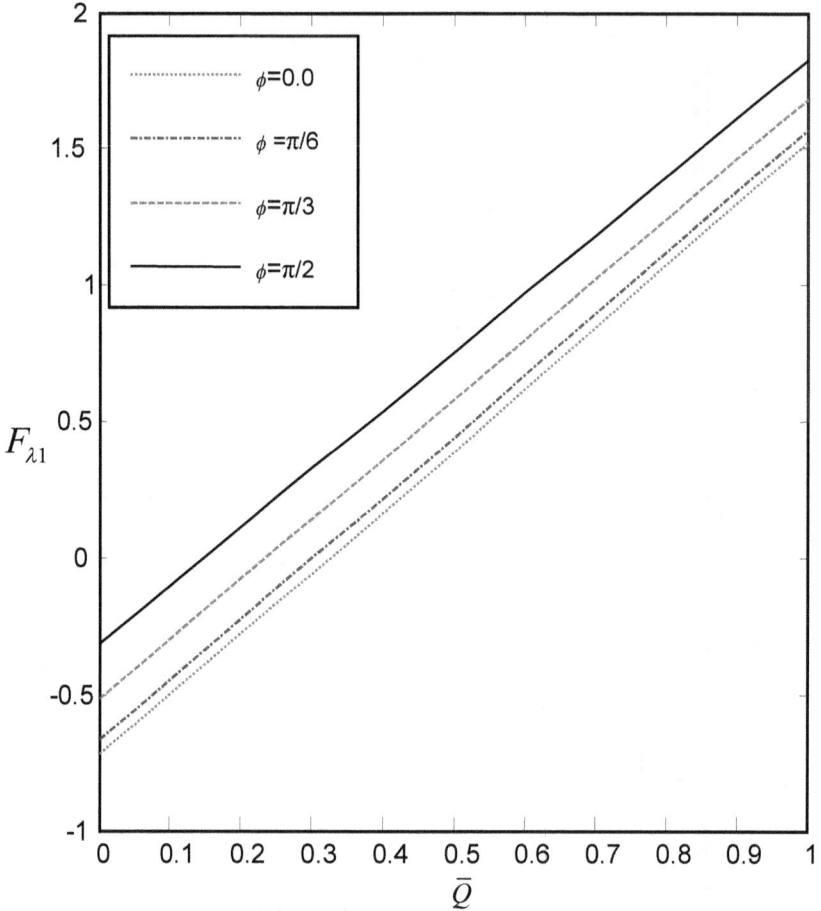

Figure 3.22: Various of frictional force $F_{\lambda 1}$ with $\bar{Q}$ at $y = h_1$ for different values of phase difference 'ϕ' with a=0.6, b=0.5, d=1, n=0.1and We =0.01.

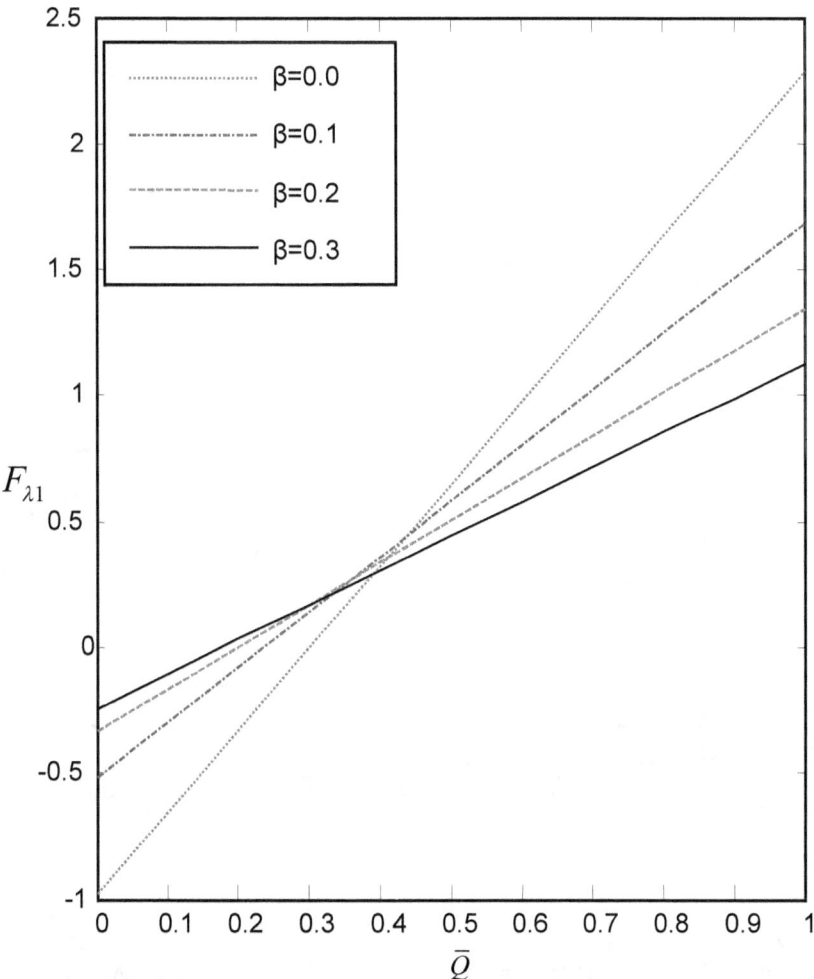

Figure 3.23: Various of frictional force $F_{\lambda1}$ with $\overline{Q}$ at $y = h_1$ for different values of permeability parameter 'β' with a=0.5, b=0.5, d=0.5, ϕ=π/6, n=0.1 and We =0.01.

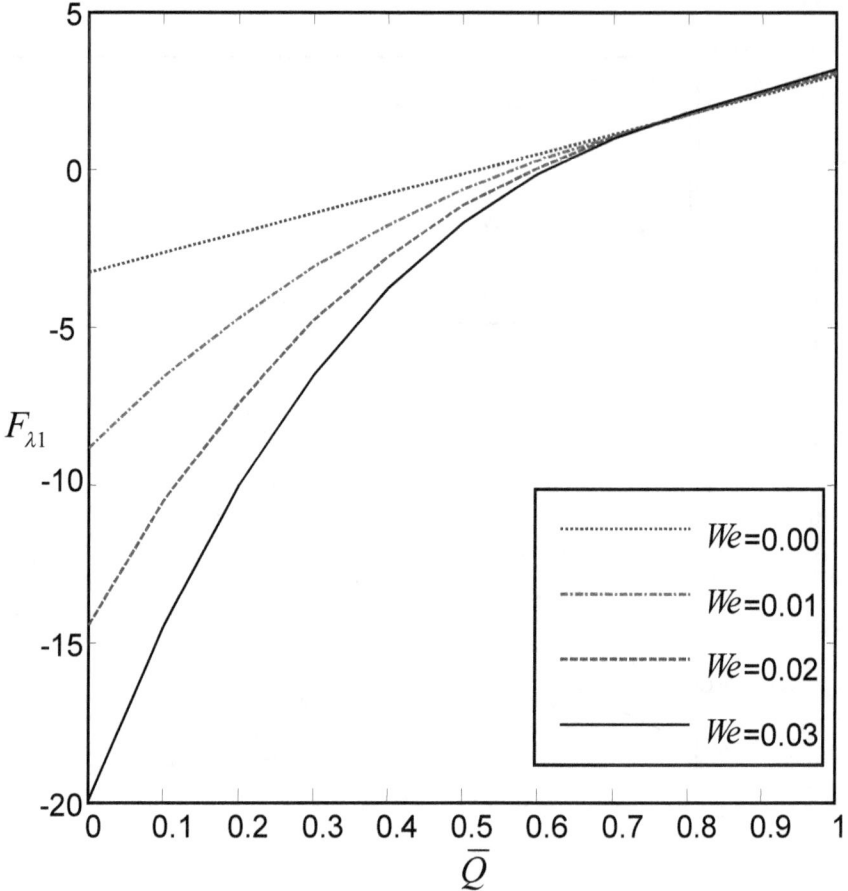

Figure 3.24: Various of frictional force $F_{\lambda 1}$ with $\bar{Q}$ at $y = h_1$ for different values of Weissenberg number 'We' with a=0.5, b=0.5, d=0.5, $\phi=\pi/6$, $\beta=0.01$ and n=0.1.

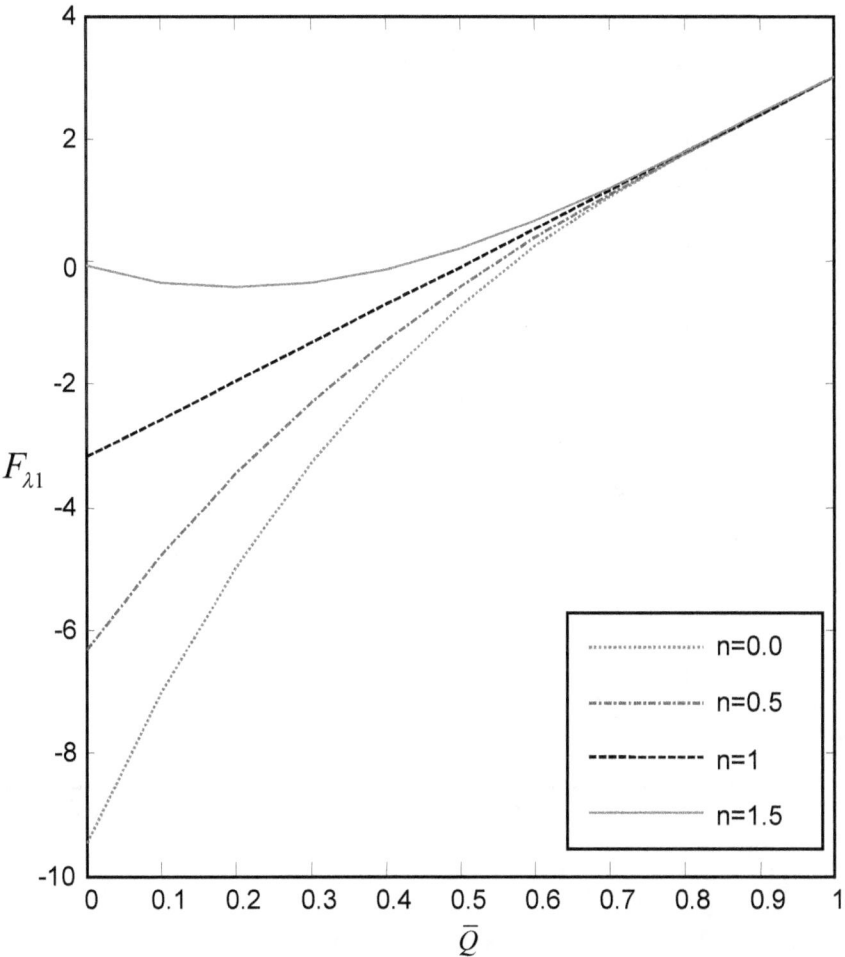

Figure 3.25: Various of frictional force $F_{\lambda 1}$ with $\overline{Q}$ at $y = h_1$ for different 'n' with

a=0.5, b=0.5, d=0.5, ϕ=π/6, β=0.01 and We =0.01.

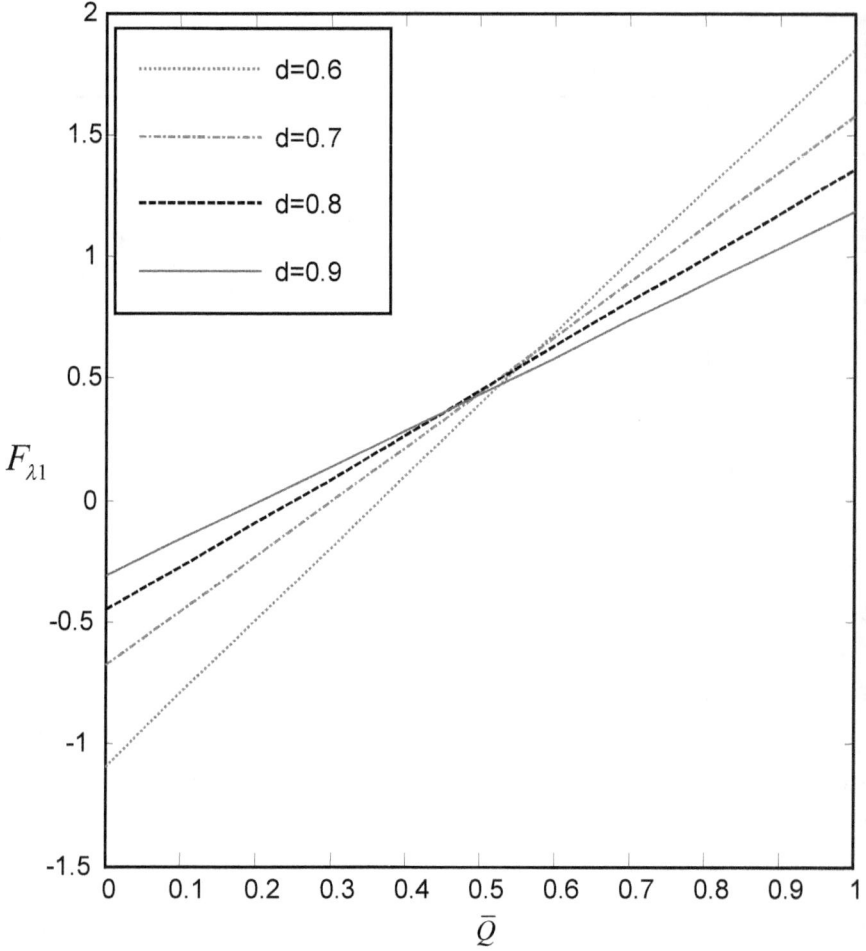

Figure 3.26: Various of frictional force $F_{\lambda 1}$ with $\overline{Q}$ at $y = h_1$ for different channel width 'd' with a=0.5, b=0.5, $\phi = \pi/6$, $\beta = 0.01$, n=0.5 and $We = 0.01$.

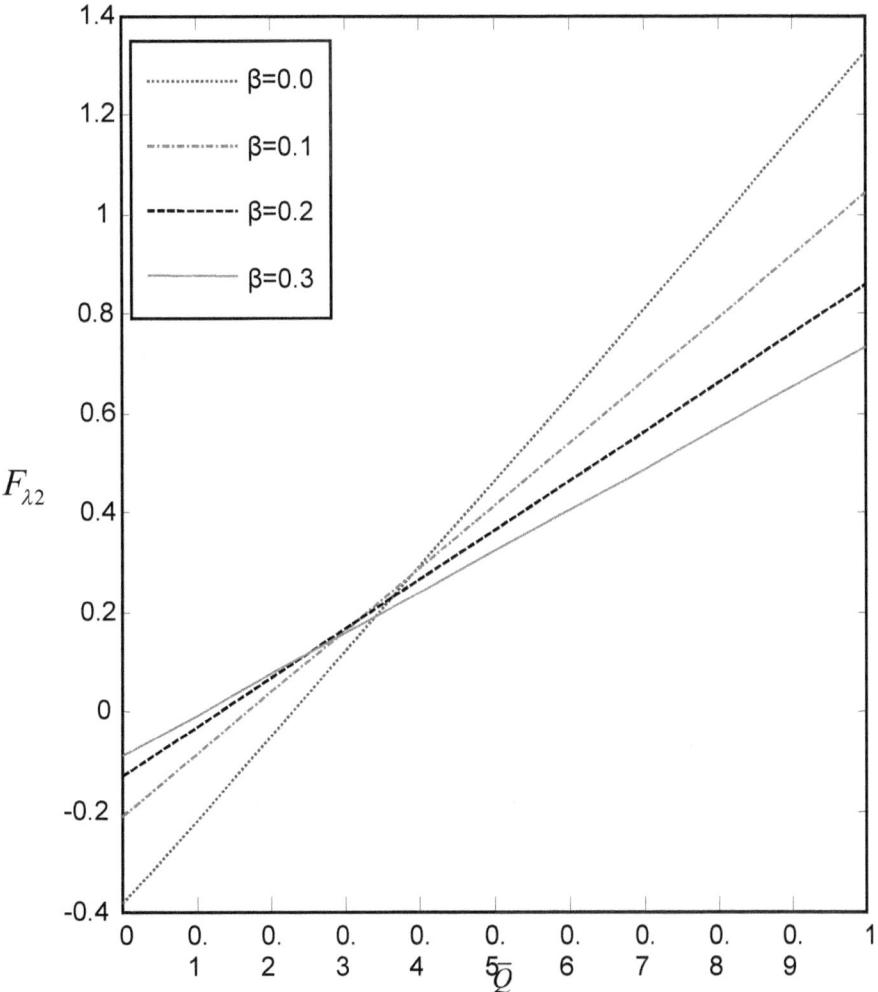

Figure 3.27: Various of frictional force $F_{\lambda2}$ with $\overline{Q}$ at $y = h_1$ for different values of permeability parameter 'β' with a=0.5, b=0.5, d=1, $\phi=\pi/6$, n=0.5 and We =0.01.

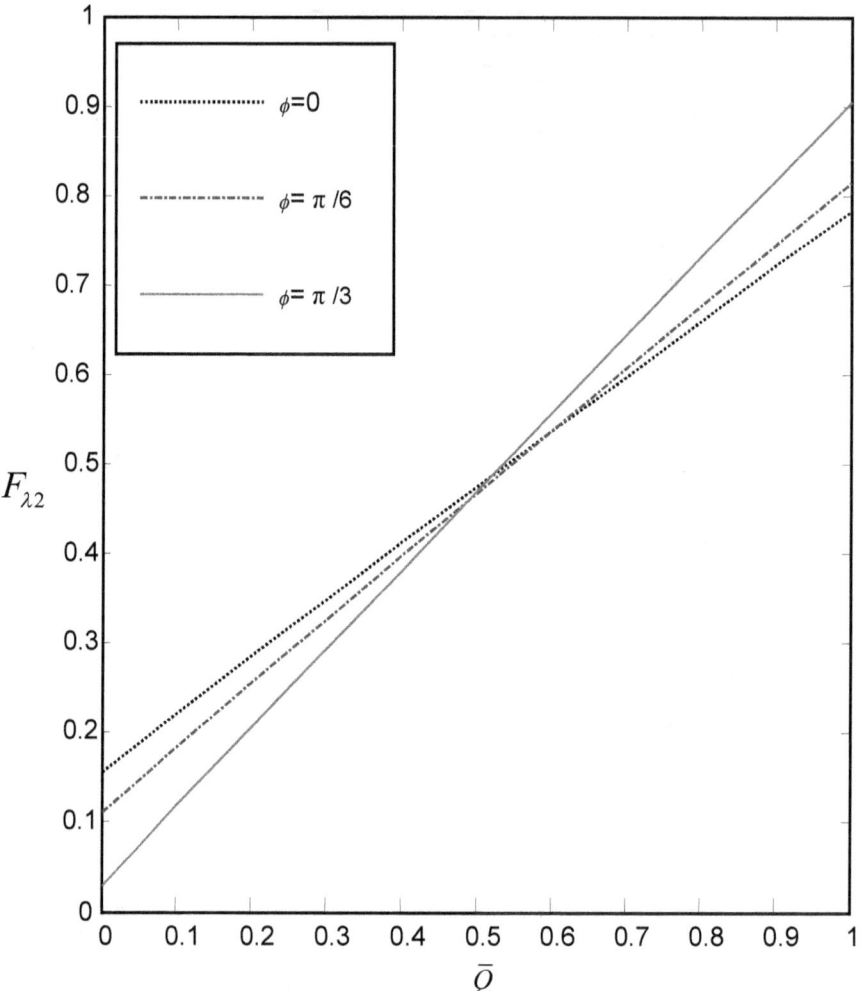

Figure 3.28: Various of frictional force $F_{\lambda 2}$ with $\overline{Q}$ at $y = h_1$ for different values of phase difference 'ϕ' with a=0.5, b=0.5, d=1, β=0.01, n=0.5 and We =0.01.π

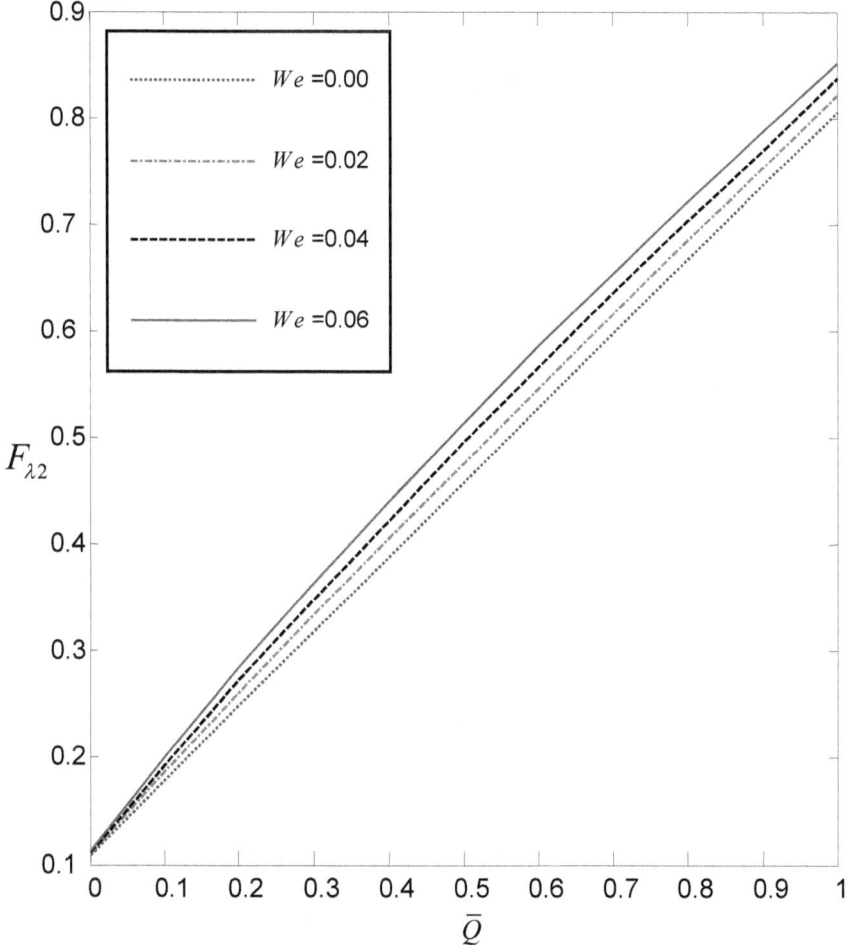

Figure 3.29: Various of frictional force $F_{\lambda 2}$ with $\overline{Q}$ at $y = h_1$ for different values of Weissenberg number 'We' with a=0.5, b=0.5, d=0.5, β=0.01, ϕ=π/6 and n=0.5.

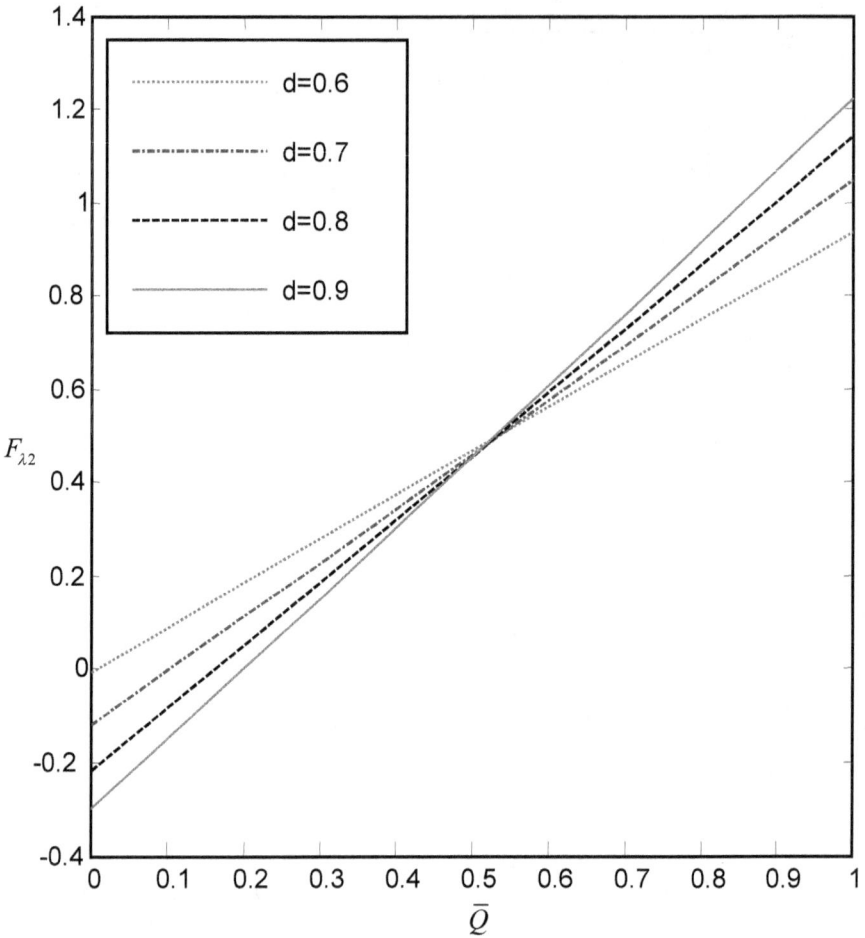

Figure 3.30: Various of frictional force $F_{\lambda 2}$ with $\bar{Q}$ at $y = h_1$ for different values of channel width 'd' with a=0.5, b=0.5, d=1, β=0.1, ϕ=π/6, n=0.5 and We =0.01.

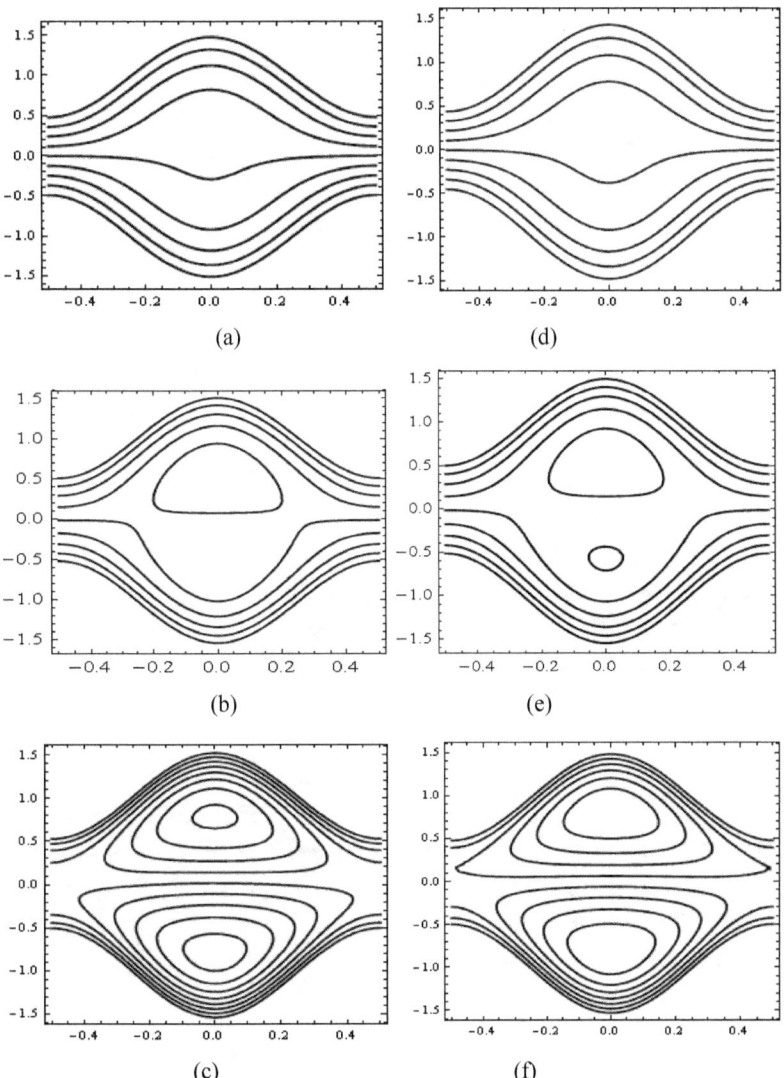

Figure 3.31: Stream lines for different values of mean flow rate $\overline{Q}$ =1(panels (a) and

(d)), $\overline{Q}$−1.4 (panels (b) and (e)) and $\overline{Q}$=1.8 (panels (c) and (f)) for Carreau

fluid (n=0.4, left panels) and Newtonian fluid (n=1, right panels). The other

parameters chosen are a=0.5, b=0.5, d=1, β=0.1, ϕ=0 □and We =0.01.

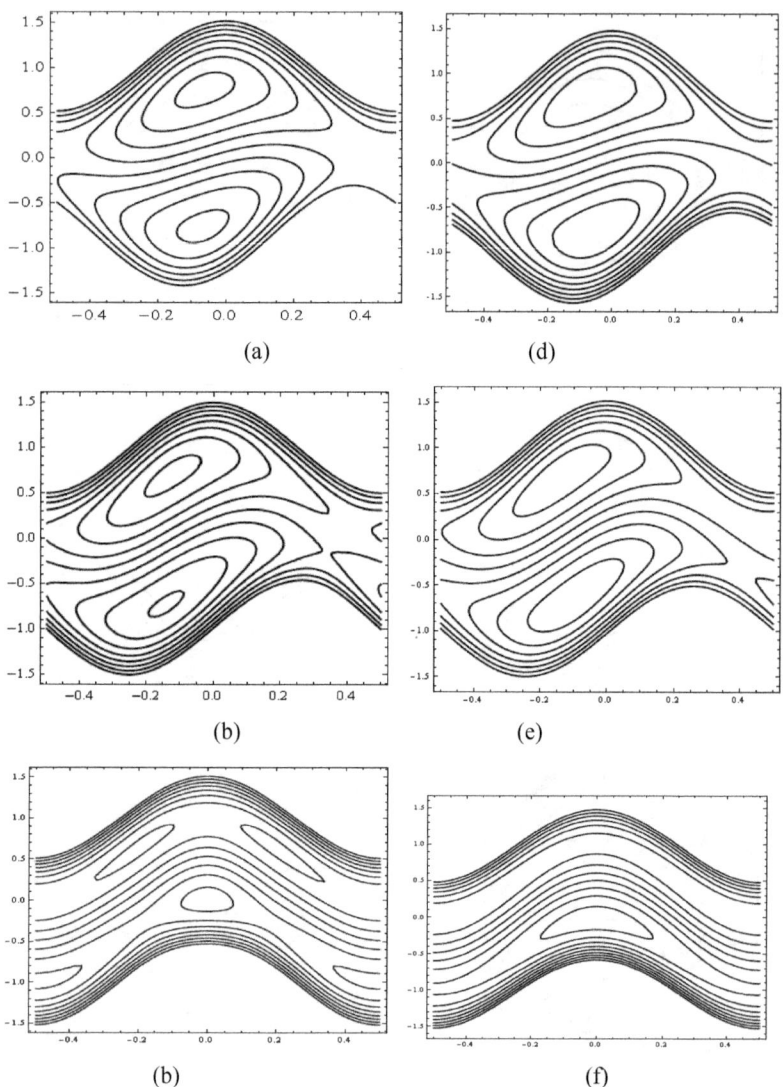

Figure 3.32: Stream lines for different phase difference $\phi = \pi/4$ (panels (a) and (d)), $\phi = \pi/2$ (panels (b) and (e)) and $\phi = \pi$ (panels (c) and (f)) for Carreau fluid (n=0.2, left panels) and Newtonian fluid (n=1, right panels). The other parameters chosen are a=0.5, b=0.5, d=1, β=0.1, $\overline{Q}$ =1.8 and We =0.01

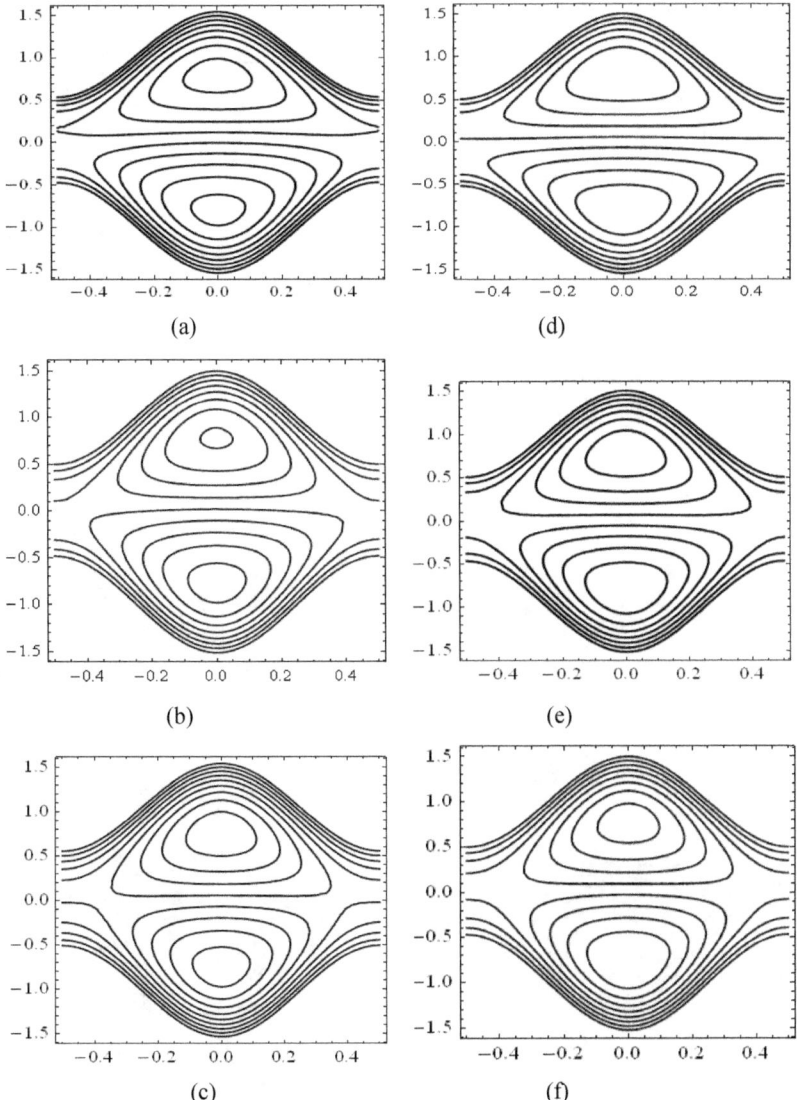

Figure 3.33: Stream lines for different permeability parameter β=0 (panels (a) and (d)), β=0.2 (panels (b) and (e)) and β=0.4 (panels (c) and (f)) for Carreau fluid (n=0.2, left panels) and Newtonian fluid (n=1, right panels). The other parameters chosen are a=0.5, b=0.5, d=1, ϕ=0, $\bar{Q}$=1.8 □and We =0.01.

147

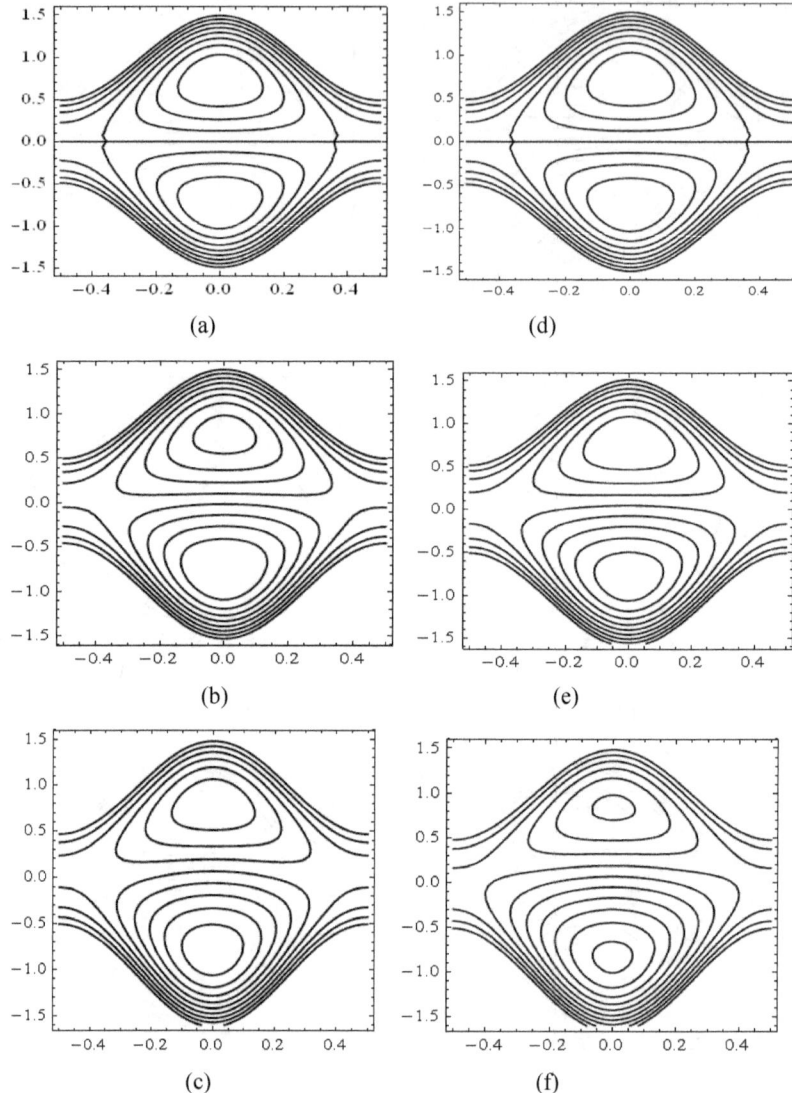

Figure 3.34: Stream lines for different Weissenberg number We =0□ (panels (a) and (d)), We =0.2 (panels (b) and (e)) and We =0.04 (panels (c) and (f)) for Carreau fluid (n=0.2, left panels) and Newtonian fluid (n=1, right panels).

The other parameters chosen are a=0.5, b=0.5, d=1, ϕ=0,□□ $\bar{Q}$=1.8 and β=0.2.

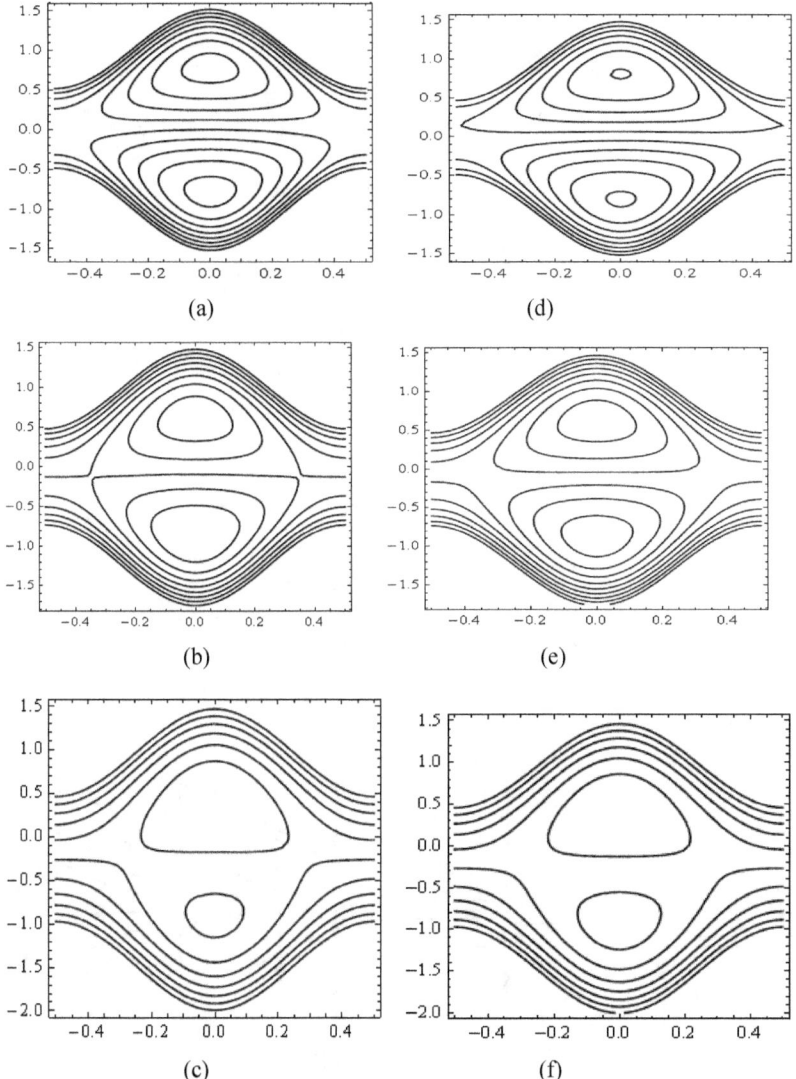

Figure 3.35: Stream lines for different channel width d=1 (panels (a) and (d)), d=1.2(panels (b) and (e)) and d=1.4 (panels (c) and (f)) for Carreau fluid (n=0.2, left panels) and Newtonian fluid (n=1, right panels). The other parameters chosen are a=0.5, b=0.5, ϕ=0, $\overline{Q}$=1.8, β=0.2 and We =0.01.

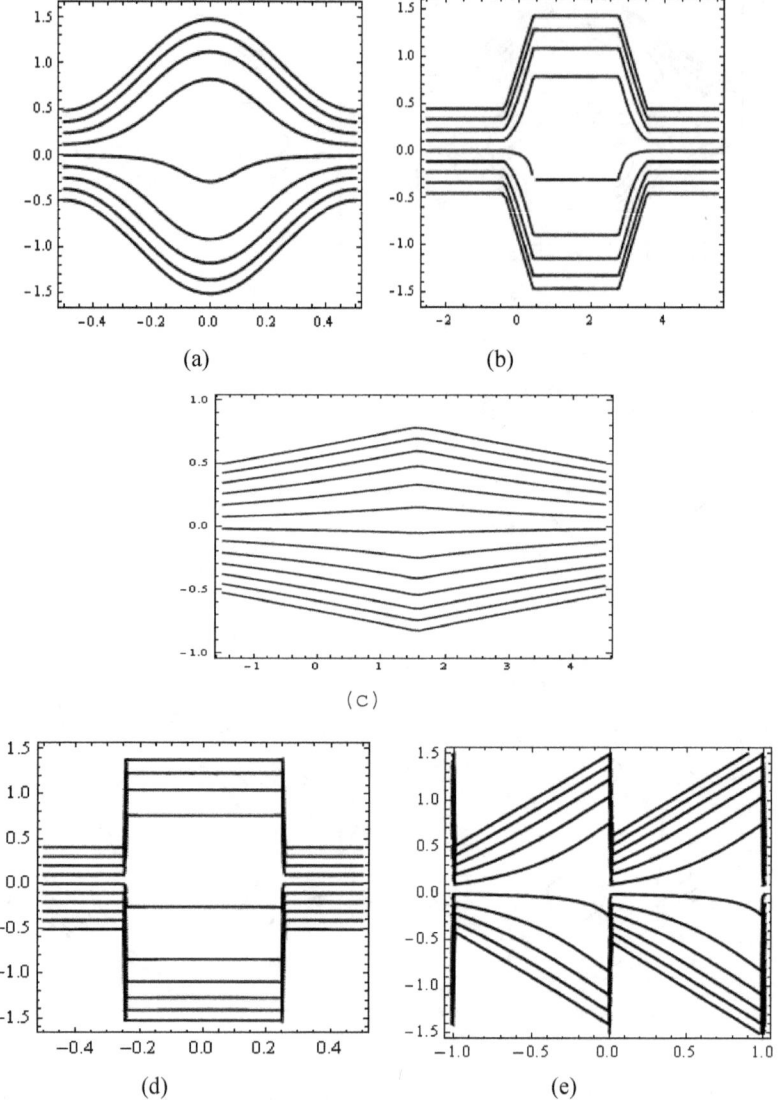

Figure 3.36: Stream lines for five different wave forms (a) sinusoidal wave,
(b) trapezoidal wave (c) triangular wave (d) square wave (e) sawtooth
wave with a=0.5, b=0.5, d=1, n=0.5, ϕ=0, $\bar{Q}$=1, β=0.2 and We =0.01.

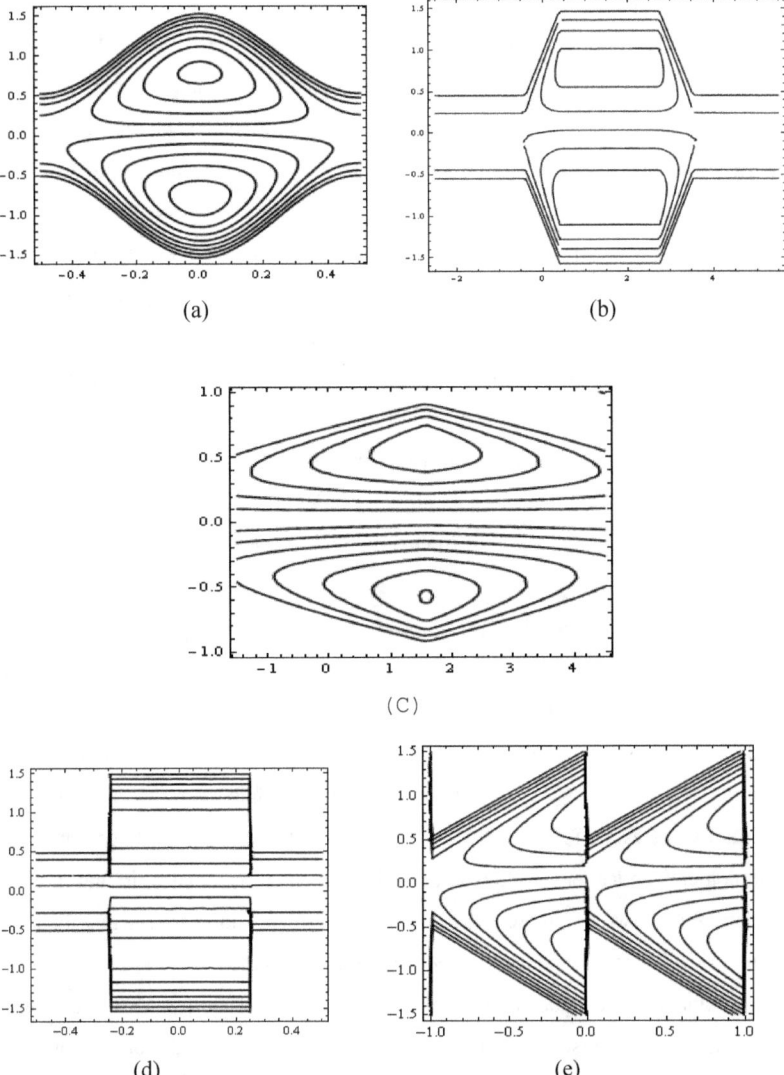

Figure 3.37: Stream lines for five different wave forms a) sinusoidal wave,
b) trapezoidal wave c) triangular wave d) square wave c) sawtooth wave with

a=0.5, b=0.5, d=1, n=0.5, ϕ=0, $\overline{Q}$=1.8,$\square$ β=0.2 and We =0.01.

4.1 Introduction

Peristaltic pumping may arise in circumstances where it is to desirable to avoid using any internal moving part such as piston in pumping process. Studying peristaltic flows, especially with a view to applications in biomechanics and physiology, one should have a complete understanding of mechanical properties of the fluid being transported representing blood vessels, intestines and ducts efferentus of the male reproductive organs and in transport of spermatozoa in the cervical cannal. In addition such mechanism has several applications in engineering and in biomechanical systems including roller and finger pumps. It has been observed from experimental investigations that blood, being a suspension of cells, behaves like a non-Newtonian fluid at low shear rates in tubes of small diameters. Non-Newtonian fluids form a broad class of fluids in which the relation connecting the shear stress and the shear rate is not linear. Therefore in order to meet the growing needs of understanding various physical phenomenon several models were developed and analyzed.

When a fluid flows over a rigid surface the usual no slip condition is valid. But in reality, when we observe the pipes manufactured it is clear that not all the inner surface of the pipe are smooth. They may be rough also. When the flow takes place over a rough/porous lined surface, the above mentioned slip condition is not valid at that surface. Further the slip condition plays an important role in shear skin, spurt and hysteresis effects. The fluids that exhibit boundary slip have important technological applications such as in polishing valves of artificial heart and internal cavities. In view of these, the investigation on flows through channels (symmetric/asymmetric) with slip effects became important.

Latham (1966) studied fluid motion in peristaltic pumps. Mishra and Rao (2004) discussed peristaltic transport of a Newtonian fluid in an asymmetric channel. Wazwaz (2005) examined a decomposition method for a reliable treatment of the Emden-Fowler equation. Hosseini abd Nasabzadeh (2006) analyzed on the convergence of Adomian decomposition method. Hayat (2006) presented the analysis of a mathematical model of peristalsis in tubes through a porous medium. Elshehaway (2006) made peristaltic transport in an asymmetric channel through a porous medium. Hayat *et al.* (2007) also discussed influence of partial slip on the peristaltic flow in a porous medium. The works so far available in asymmetric channels do not deal with the effects magnetic field while discussing the several physiological/industrial situations.

In this chapter, peristaltic transport of a conducting fluid in an asymmetric channel through a porous medium is investigated. The velocity field, the pressure rise, the frictional force are determined and the results are deduced and discussed.

4.2 Mathematical formulation

Consider the two-dimensional channel of width $d_1 + d_2$ (Figure 4.1) filled with an incompressible conducting viscous fluid. The asymmetry in the channel flow is induced due to different amplitudes and phases of the peristaltic waves on the channel walls. The geometry of the wall surfaces is given by

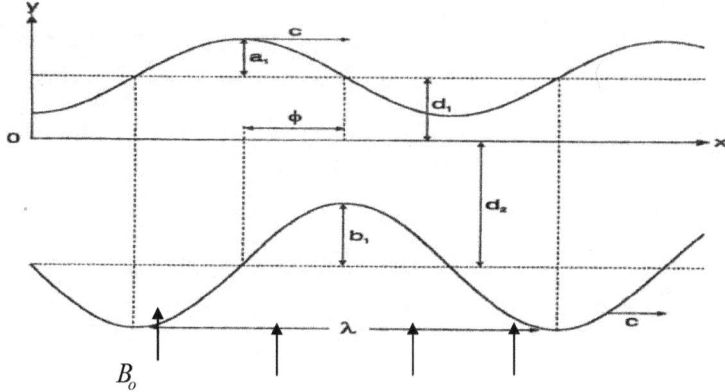

Figure 4.1.Schematic diagram of a two-dimensional asymmetric channel

$$H_1 = d_1 + a_1 \cos\left[\frac{2\pi}{\lambda}(\bar{X} - c\bar{t})\right] \dots\dots \text{ upper wall,} \qquad (4.1)$$

$$H_2 = -d_2 - b_1 \cos\left[\frac{2\pi}{\lambda}(\bar{X} - c\bar{t}) + \phi\right] \dots\dots\text{lower wall.} \qquad (4.2)$$

where a_1 and b_1 are the amplitudes of the waves, λ is the wavelength and $\phi(0 \le \phi \le \pi)$ is the phase difference. It is noted that $\phi = 0$ corresponds to the symmetric channel case with waves out of phase and for $\phi = \pi$ waves are in phase. Furthermore, a_1, b_1, d_1, d_2 and ϕ satisfy the following condition

$$a_1^2 + b_1^2 + 2a_1b_1 \cos \phi \le (d_1 + d_2)^2 . \tag{4.3}$$

4.3 Equations of motion

The corresponding flow equations in a porous medium are

$$\frac{\partial \bar{U}}{\partial \bar{X}} + \frac{\partial \bar{V}}{\partial \bar{Y}} = 0, \tag{4.4}$$

$$\frac{\partial \bar{U}}{\partial \bar{X}} + \bar{U}\frac{\partial \bar{U}}{\partial \bar{X}} + \bar{V}\frac{\partial \bar{U}}{\partial \bar{Y}} = -\frac{1}{\rho}\frac{\partial \bar{P}}{\partial \bar{X}} + v\left(\frac{\partial^2 \bar{U}}{\partial \bar{X}^2} + \frac{\partial^2 \bar{U}}{\partial \bar{Y}^2}\right) - \frac{v}{K}\bar{U} - \sigma B_0^2 \bar{U} , \tag{4.5}$$

$$\frac{\partial \bar{V}}{\partial \bar{t}} + \bar{U}\frac{\partial \bar{V}}{\partial \bar{X}} + \bar{V}\frac{\partial \bar{V}}{\partial \bar{Y}} = -\frac{1}{\rho}\frac{\partial \bar{P}}{\partial \bar{X}} + v\left(\frac{\partial^2 \bar{V}}{\partial \bar{X}^2} + \frac{\partial^2 \bar{V}}{\partial \bar{Y}^2}\right) - \frac{v}{K}\bar{V}, \tag{4.6}$$

In the above equations $\bar{U}$ and $\bar{V}$ are the respective velocity components in the $\bar{X}$ and $\bar{Y}$ directions of the fixed frame, ρ is the constant density of the fluid, $\bar{P}$ is the fluid pressure, v is the kinematic viscosity and K is permeability of the porous medium. Bars indicate that the respective quantity has physical dimension. The flow is inherently unsteady in the fixed frame but it can be treated as steady by switching from the fixed frame to the wave frame. If $\bar{u}$ and $\bar{v}$ denote the velocity components in the $\bar{x}$ and $\bar{y}$ directions of the wave frame, then

$$\bar{x} = \bar{X} - ct , \ \bar{y} = \bar{Y} , \ \bar{u} = \bar{U} - c , \bar{v} = \bar{V} , \ \ \bar{p}(x) = \bar{P}(\bar{X}, \bar{t}) , \tag{4.7}$$

where c is the speed of the peristaltic wave.

We introduce the following dimensionless quantities

$$x = \frac{\bar{X}}{\lambda} , \ y = \frac{\bar{y}}{d_1} , \ u = \frac{\bar{u}}{c} , \ v = \frac{\bar{v}}{c\delta} \ , \ \delta = \frac{d_1}{\lambda} , \ d = \frac{d^2}{d_1} , \ p = \frac{d_1^2 \bar{p}}{\mu c \lambda} , \ t = \frac{c\bar{t}}{\lambda} , \ h_1 = \frac{H_1}{d_1}$$

$$, \ h_2 = \frac{H_2}{d_1} , \ a = \frac{a_1}{a_1} , \ b = \frac{b_1}{d_1} , \ Da = \frac{\bar{K}}{d_1^2} , \ R_e = \frac{cd_1}{\mu} , \ M^2 = \frac{\sigma B_0^2 d_1^2}{v} .$$

The resulting flow equation in terms of the stream function ψ

$$\left(u = \frac{\partial \psi}{\partial y}, \ v = -\frac{\partial \psi}{\partial x}\right) \text{is}$$

$$R_e \delta \left[\psi_y \psi_{yyx} - \psi_x \psi_{yyy} + \delta^2 (\psi_y \psi_{xxx} - \psi_y \psi_{xxy}) \right]$$

$$= \psi_{yyyy} + 2\delta^2 \psi_{xxyy} + \delta^4 \psi_{xxxx} - \left(\frac{1}{Da} + M^2 \right) \psi_{yy} - \frac{\delta^2}{K} \psi_{xx}. \qquad (4.8)$$

where the subscripts x and y indicate partial differentiations with respect to variables x and y respectively.

The dimensionless flux in the fixed frame is defined by

$$Q = \int_{h_2}^{h_1} (u+1)dy = \int_{h_2}^{h_1} u\,dy + \int_{h_2}^{h_1} dy = q + h_1 - h_2, \qquad (4.9)$$

The average volume flow rate over one period $\left(T = \dfrac{\lambda}{c} \right)$ of the peristaltic wave

is

$$\overline{Q} = \frac{1}{T} \int_0^T Q\,dt = \frac{1}{T} \int_0^T (q + h_1 - h_2)dt = q + 1 + d, \qquad (4.10)$$

The dimensionless boundary conditions express that the fluid does not detach from the walls and that the slip velocity is proportional to the bounding shear traction. In terms of the stream function they are

$$\psi = \frac{q}{2} \ at \ y = h_1(x) = 1 + a \cos 2\pi x, \qquad (4.11)$$

$$\psi = -\frac{q}{2} \ at \ y = h_2(x) = -d - b\cos(2\pi x + \phi), \qquad (4.12)$$

$$\frac{\partial \psi}{\partial y} + L \frac{\partial^2 \psi}{\partial y^2} = -1 \ at \ y = h_1(x), \qquad (4.13)$$

$$\frac{\partial \psi}{\partial y} - L \frac{\partial^2 \psi}{\partial y^2} = -1 \ at \ y = h_2(x). \qquad (4.14)$$

in which L is the non-dimensional permeability parameter, q is the flux in wave frame and a, b, ϕ and d satisfy the relation

$$a^2 + b^2 + 2ab\cos\phi \le (1+d)^2.$$

We note that the partial slip conditions (4.13) and (4.14) are applied at the walls following Nadeem and Akram (2010).

155

Adopting the long wavelength $\delta \ll 1$ and low Reynolds number approximations equation (4.8) takes the limiting form

$$\psi_{yyyy} - \left(\frac{1}{Da} + M^2\right)\psi_{yy} = 0,$$
(4.15)

The mathematical consisting of equations (4.11) to (4.12) are be solved by the Adomian decomposition method.

4.4 Solution of the problem

In order to the Adomian decomposition method, equation (4.15) can be written as

$$\psi = \left(\frac{1}{A}\right)\mathcal{L}^{-1}\psi_{yy},$$
(4.16)

where $\frac{1}{A} = \frac{1}{Da} + M^2$ and $\mathcal{L} = \frac{d^4}{dy^4}$. Since a fourth-order difference operator, $\mathcal{L}^{-1}$ is a fourth-fold integration operator defined by

$$\mathcal{L}^{-1} = \int_0^y \int_0^\tau \int_0^\eta \int_0^\xi (.)\,d\xi\,d\eta\,d\tau\,dy.$$
(4.17)

Operating with $\mathcal{L}^{-1}$, equation (4.17) yields

$$\psi = c_1 + c_2 y + c_3 \frac{y^2}{2!} + c_4 \frac{y3}{3!} + \mathcal{L}^{-1}(\psi_{yy}).$$
(4.18)

in which the function c_i (i=1 to 4) can be determine by utilizing the boundary conditions (4.11) to (4.14).

By the standard Adomian decomposition method, one can write

$$\psi = \sum_{m=0}^{\infty} \psi_m.$$
(4.19)

where the components $\psi_m, m \geq 0$, will be determine recursively. From equations (4.17) and (4.18), we obtain the following recursively relation

$$\psi_0 = C_1 + C_2 y + C_3 \frac{y^2}{2!} + C_4 \frac{y3}{3!},$$
(4.20)

$$\psi_{m+1} = \left(\frac{1}{A}\right)\mathcal{L}^{-1}(\psi_{yy}), \quad m \geq 0.$$
(4.21)

156

whence

$$\psi_1 = AC_3 \frac{\left(\frac{y}{\sqrt{A}}\right)^4}{4!} + A\sqrt{A}C_4 \frac{\left(\frac{y}{\sqrt{A}}\right)^5}{5!} ,$$

$$\psi_2 = AC_3 \frac{\left(\frac{y}{\sqrt{A}}\right)^6}{6!} + A\sqrt{A}C_4 \frac{\left(\frac{y}{\sqrt{A}}\right)^7}{7!} ,$$

$$\psi_m = AC_3 \frac{\left(\frac{y}{\sqrt{A}}\right)^{2m+2}}{(2m+1)!} + A\sqrt{A}C_4 \frac{\left(\frac{y}{\sqrt{A}}\right)^{2m+3}}{(2m+3)!} , \quad m \geq 0.$$

Through equation (4.19), the expression for ψ is easily seen to have the form

$$\psi = C_1 + C_2 y + AC_3 \left(\cosh \frac{y}{\sqrt{A}} - 1 \right) + A\sqrt{A}C_4 \left(\sinh \frac{y}{\sqrt{A}} - \frac{y}{\sqrt{A}} \right). \qquad (4.22)$$

which may be simplified as,

$$\psi = F_1 + F_2 y + F_3 \cosh \frac{y}{\sqrt{A}} + F_4 \sinh \frac{y}{\sqrt{A}} . \qquad (4.23)$$

The velocity is given

$$u = F_2 + \frac{1}{\sqrt{A}} F_3 \sinh \frac{y}{\sqrt{A}} + \frac{1}{\sqrt{A}} F_4 \cos \frac{y}{\sqrt{A}} . \qquad (4.24)$$

where the values of F_1 to F_4 can be found by using the boundary conditions (4.11) to (4.14) are given by

$$F_1 = \frac{-(h_1 - h_2)\left(q + \left(\frac{q\beta + 2A}{\sqrt{A}} \right) \tanh\left[\frac{h_1 - h_2}{2\sqrt{A}} \right] \right)}{2(h_1 - h_2) - 2\left(\frac{\beta(h_2 - h_1) + 2A}{\sqrt{A}} \right) \tanh\left[\frac{h_1 - h_2}{2\sqrt{A}} \right]} ,$$

$$F_2 = \frac{q + \left(\frac{q\beta + 2A}{\sqrt{A}} \right) \tanh\left[\frac{h_1 - h_2}{2\sqrt{A}} \right]}{(h_1 - h_2) - \left(\frac{\beta(h_2 - h_1) + 2A}{\sqrt{A}} \right) \tanh\left[\frac{h_1 - h_2}{2\sqrt{A}} \right]} ,$$

$$F_3 = \frac{-\sqrt{A}\left(h_1 - h_2 + q\right)\operatorname{sech}\left[\dfrac{h_1 - h_2}{2\sqrt{A}}\right]\sinh\left[\dfrac{h_1 + h_2}{2\sqrt{A}}\right]}{\left(h_2 - h_1\right) + \left(\dfrac{\beta\left(h_2 - h_1\right) + 2A}{\sqrt{A}}\right)\tanh\left[\dfrac{h_1 - h_2}{2\sqrt{A}}\right]},$$

$$F_4 = \frac{\sqrt{A}\left(h_1 - h_2 + q\right)\operatorname{sech}\left[\dfrac{h_1 - h_2}{2\sqrt{A}}\right]\cosh\left[\dfrac{h_1 + h_2}{2\sqrt{A}}\right]}{\left(h_2 - h_1\right) + \left(\dfrac{\beta\left(h_2 - h_1\right) + 2A}{\sqrt{A}}\right)\tanh\left[\dfrac{h_1 - h_2}{2\sqrt{A}}\right]}.$$

The resulting non-dimensional form equation (4.5) for the axial pressure gradient under the long wavelength assumption is

$$\frac{dp}{dx} = \psi_{yyy} - \frac{1}{A}\psi_y - \frac{1}{A}$$

$$= \frac{\left(h_1 - h_2 + q\right)\left(\sqrt{A} - \beta\tanh\left[\dfrac{h_1 - h_2}{2\sqrt{A}}\right]\right)}{A^{\frac{3}{2}}\left(h_2 - h_1\right) + A\left(2A + \left(h_2 - h_1\right)\right)\beta\tanh\left[\dfrac{h_1 - h_2}{2\sqrt{A}}\right]}. \qquad (4.25)$$

The pressure rise per wavelength is

$$\Delta p = \int_0^1 \frac{dp}{dx}\,dx. \qquad (4.26)$$

The frictional force at $y = h_1$ and $y = h_2$ denoted by $F_{\lambda 1}$ and $F_{\lambda 2}$ respectively are given by as follows

$$F_{\lambda 1} = \int_0^1 -h_1^2\left(\frac{dp}{dx}\right)dx, \qquad (4.27)$$

$$F_{\lambda 2} = \int_0^1 -h_2^2\left(\frac{dp}{dx}\right)dx. \qquad (4.28)$$

4.5 Expressions for wave shape:

The non-dimensional expressions for three considered wave forms are given by the following equations:

1. Sinusoidal wave

$$h(x) = 1 + a\sin(x),\tag{4.29}$$

2. Triangular wave

$$h(x) = 1 + a\left\{\frac{8}{\pi^3}\sum_{m=1}^{\infty}\frac{(-1)^{m+1}}{(2m-1)^2}\sin[(2m-1)x]\right\},\tag{4.30}$$

3. Trapezoidal wave

$$h(x) = 1 + a\left\{\frac{32}{\pi^2}\sum_{m=1}^{\infty}\frac{\sin\frac{\pi}{8}(2m-1)}{(2m-1)^2}\sin[(2m-1)x]\right\}.\tag{4.31}$$

4.6 Results and discussion:

The variation of velocity u with y is calculated from equation (4.24) for different values of Hartmann number M with x=1, a=0.5, b=0.5, d=1, $\phi = \frac{\pi}{6}$, β =0.1, $\overline{Q}$ =1 and Da =0.1 and is depicted in figure (4.2). It is observed that the velocity u increases with increasing Hartmann number M near the walls. However, u decreases with increasing M near the centre of the channel.

The relationship between velocity u with y is plotted in figure (4.3) for different values of the permeability parameter β with x=1, a=0.5, b=0.5, d=1, M=1, $\phi = \frac{\pi}{3}$, $\overline{Q}$ =1, and Da =0.1. It is observed that the velocity u increases with increasing permeability parameter β near the walls. However, u decreases by increasing β near the centre of the channel.

The variation between velocity u with y is shown in figure (4.4) for different values of the phase difference ϕ with x=1, a=0.5, b=0.5, d=1, $\overline{Q}$ =2, β =0.1, M=1 and Da =0.1. It is concluded that the velocity decreases with increasing phase difference ϕ in the lower half of the channel.

In figure (4.5) the relation between velocity u with y is plotted for different values of the average volume flow rate $\overline{Q}$ with x=1, a=0.5, b=0.5, d=1, $\phi = \frac{\pi}{3}$, M=1, Da =0.1 and β =0.1. It is observed that the velocity increases with increasing the average volume flow rate $\overline{Q}$.

The relationship between velocity u with y is shown for different values of the Darcy Da with x=1, a=0.5, b=0.5, d=1, M=1, $\phi = \frac{\pi}{6}$, $\overline{Q}$=1 and β=0.1 and in figure (4.6). It is observed that the velocity u increases with increasing Darcy number Da near the walls. However, u decreases by increasing Da near the centre of the channel.

The variation between velocity u with y is drawn in figure (4.7) for different values of the channel width d with x=1, a=0.5, b=0.5, $\overline{Q}$=2, β =0.1, M=1, $\phi = \frac{\pi}{6}$ and Da =0.1. It is concluded that the velocity decreases in upper half of the channel and increases lower half of the channel with increasing channel width d.

The variation of pressure gradient $\frac{dp}{dx}$ with x is calculated from equation (4.25) for different values of Hartmann number with a=0.6, b=0.6, d=1, ϕ=0, $\overline{Q}$=1, β =0.1 and Da =0.1 and is plotted in figure (4.8). It is observed that pressure gradient $\frac{dp}{dx}$ increases with increasing Hartmann number M.

The variation between pressure gradient $\frac{dp}{dx}$ with x is depicted in figure (4.9) for different values of permeability parameter β with a=0.6, b=0.6, d=1, M=1, $\overline{Q}$=1, ϕ=0 and Da =0.1. It is concluded that the pressure gradient $\frac{dp}{dx}$ increases with decreasing permeability parameter β.

The variation of pressure gradient $\frac{dp}{dx}$ with x for difference values of phase difference ϕ with a=0.6, b=0.6, d=1, $\overline{Q}$=1, β =0.01, M=1 and Da =0.1 and is plotted in figure (4.10). It is observed that pressure gradient $\frac{dp}{dx}$ decreases with the increase in the phase difference ϕ and point of maximum amplitudes decreases with increasing ϕ.

In figure (4.11) the relation between pressure gradient $\frac{dp}{dx}$ and x is plotted at different values of the mean flux $\overline{Q}$ with a=0.6, b=0.6, d=1, M=1, ϕ=0, β=0.01 and

160

Da =0.1. It is noticed that the pressure gradient $\frac{dp}{dx}$ increases with increasing mean

flux $\overline{Q}$.

From figure (4.12) the variation pressure gradient $\frac{dp}{dx}$ with x is plotted for

different channel width d with a=0.6, b=0.6, $\overline{Q}$=1, ϕ=0, β=0.01 and Da =0.1. It is

concluded that the pressure rise deceases with increasing channel width d.

The variation on pressure rise Δp_λ with mean flux $\overline{Q}$ is calculated from

equation (4.26) and is drawn figure (4.13) for different values of Hartmann number

M with a=0.5, b=0.5, d=1, $\phi = \frac{\pi}{6}$, β =0.01 and Da =0.1. We conclude that for values

of $\overline{Q}$ between 0.5 and 0.6, the pumping curves intersect at a point (0.55, 0). For a

given mean flux $\overline{Q}$, the pressure rise Δp_λ increases with decreasing Hartmann

number M below this point and opposite behaviour is observed above this point.

The relation between pressure rise ΔP_λ with mean flux $\overline{Q}$ is plotted figure

(4.14) for different values of phase difference ϕ with a=0.6, b=0.6, d=0.5, β =0.01,

M=1 and Da =0.1. We observe that for values of $\overline{Q}$ between 1.2 and 1.4, the

pumping curves intersect at a point (1.38, -5.5). For a given mean flux $\overline{Q}$, the

pressure rise ΔP_λ increases with increasing phase difference ϕ above this point and

opposite behaviour is observed below this point.

The relation between pressure rise ΔP_λ with mean flux $\overline{Q}$ is depicted figure

(4.15) for different values of permeability parameter β with a=0.7, b=0.7, d=1,

M=1, $\phi = \frac{\pi}{6}$ and Da =0.1. We observe that for values of $\overline{Q}$ between 0.4 and 0.5, the

pumping curves intersect at a point (0.42, 0.5). For a given mean flux $\overline{Q}$, the

pressure rise ΔP_λ increases with increasing permeability parameter β above this

point and opposite behaviour is observed below this point.

Figure (4.16) shows the relation between pressure rise ΔP_λ with mean flux $\overline{Q}$ for various values of Darcy number Da with a=0.7, b=0.7, d=1, $\phi = \dfrac{\pi}{6}$, M=1 and β =0.01. We observe that for values of $\overline{Q}$ between 0.8 and 0.9, the pumping curves intersect at a point (0.86, 0). For a given mean flux $\overline{Q}$, the pressure rise ΔP_λ increases with decreasing Darcy number Da above this point and opposite behaviour is observed below this point.

The relationship between pressure rise ΔP_λ with mean flux $\overline{Q}$ is drawn figure (4.17) for different values of channel width d with a=0.7, b=0.7, $\phi = \dfrac{\pi}{6}$, M=1, β =0.01 and Da =0.1. We conclude that for values of $\overline{Q}$ between 1 and 1.2, the pumping curves intersect at a point (1.05, -4). For a given flux $\overline{Q}$, the ϕpressure rise ΔP_λ increases with increasing channel width d below this point and opposite behaviour is observed above this point.

The variation of frictional forces $F_{\lambda 1}$ and $F_{\lambda 2}$ with $\overline{Q}$ is shown figures (4.18) to (4.27). We observe that the frictional force shows opposite behaviour to that of pressure rise for the corresponding variations in the physical parameters ϕ, β, M, d and Da.

4.6.1 Trapping phenomena

Another interesting phenomenon in peristaltic motion is trapping. It is basically the formation of an internally circulating bolus of fluid by closed stream lines. The trapped bolus will be pushed ahead along the peristaltic waves.

The stream lines are depicted in figure (4.28) for different values of Hartmann number M with a=0.5, b=0.5, d=1, β=0.01, $\overline{Q}$=1.8, ϕ=0 and Da =0.1. It is noticed that the size of the trapping bolus decreases with increasing Hartmann number M and bolus disappears for M=3.

The stream lines are drawn in figure (4.29) for different values of the average flow rate $\overline{Q}$ with a=0.5, b=0.5, d=1, β=0.01, ϕ=0, M=1 and Da =0.2. It is noticed that the size of the trapping bolus increases with increasing the average flow rate $\overline{Q}$ and bolus disappears for $\overline{Q}=1.2$.

The stream lines are drawn for various values of the permeability parameter β and are shown in figure (4.30) with a=0.5, b=0.5, d=1, ϕ=0, M=1, $\overline{Q}=1.8$ and Da =0.2. It is observed that the volume of the trapping bolus decreases by increasing permeability parameter β and the trapping bolus disappears for β=0.6.

The stream lines are plotted (4.31) for various values of the Darcy number Da with a=0.5, b=0.5, d=1, ϕ=0, M=1, $\overline{Q}=1.8$ and β=0.1. It is observed that the volume of the trapping bolus increases by increasing Darcy number Da and the trapping bolus disappears for Da =0.01.

The stream lines are plotted in figure (4.32) for various values of the phase difference ϕ with a=0.5, b=0.5, d=1, β=0.1, $\overline{Q}=1.8$, M=1 and Da =0.3. It is concluded that the size of the trapping bolus moves towards left and decreases in volume as ϕ increases. For ϕ=π the bolus disappears and stream lines are parallel to the boundary walls.

The stream lines are depicted in the wave frame for pumping for different values of channel width d in figure (4.33) with a=0.5, b=0.5, β=0.1, ϕ=0, $\overline{Q}=1.8$ and Da =0.3. It is observed that the size of the trapping bolus decreases with increasing channel width d and the bolus disappears for d=1.6.

Stream lines are plotted in figures (4.34) and (4.35) for different values of $\overline{Q}$ for trapezoidal and triangular wave forms. We find that the increase in the average volume flow rate $\overline{Q}$ enhance the size of the bolus for both wave forms.

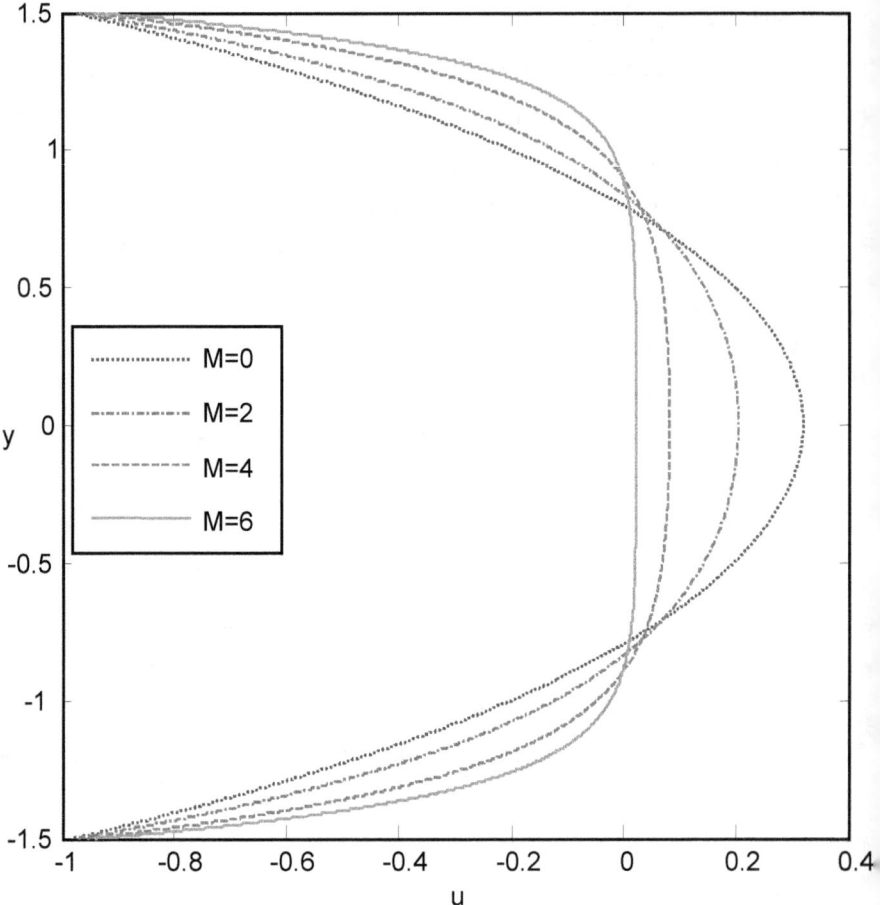

Figure 4.2: The velocity profiles with a=0.5, b=0.5, d=1, x=1, $\phi=\pi/6$,

β =0.1, $\overline{Q}$ =1 and Da =0.1.

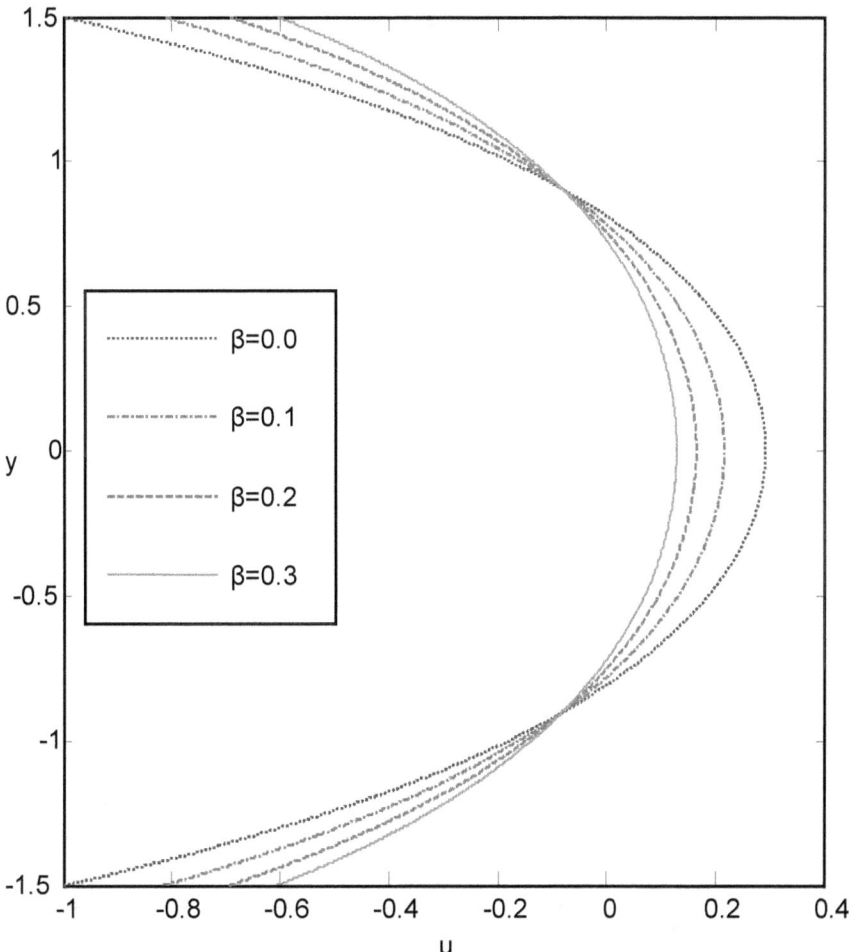

Figure 4.3: The velocity profiles with a=0.5, b=0.5, d=1, x=1, $\phi=\pi/3$, M=1, β =0.1
and Da =0.1.

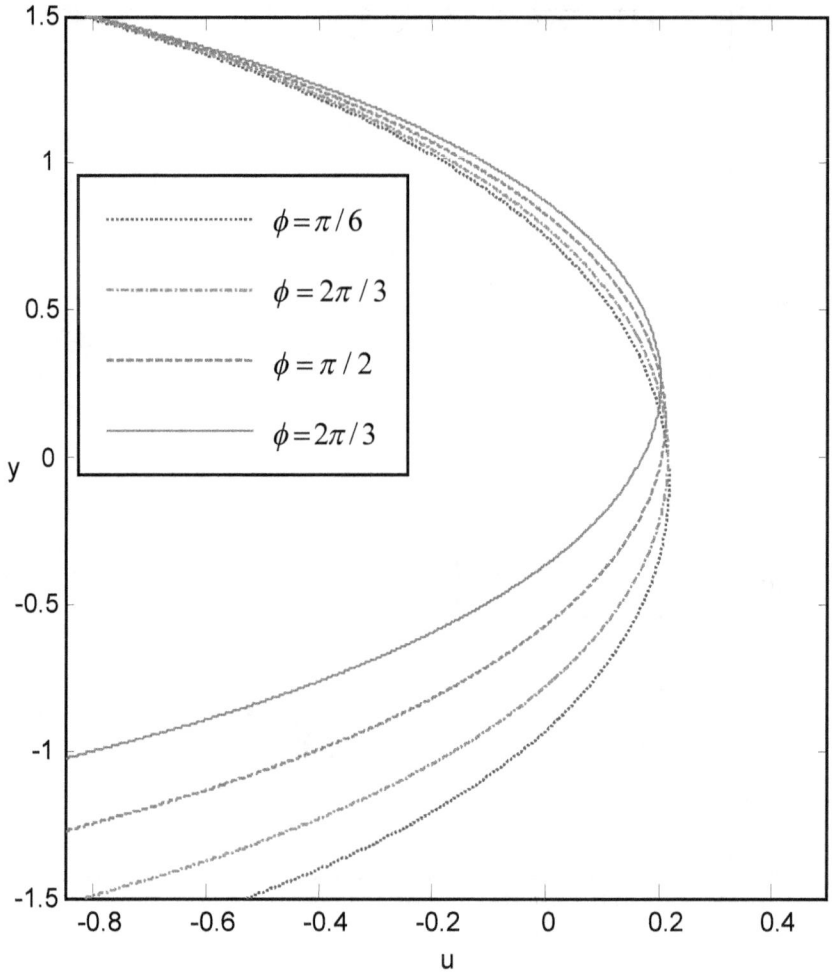

Figure 4.4: The velocity profiles with a=0.5, b=0.5, d=1, x=1, M=1,

β =0.1, $\overline{Q}$ =2 and Da =0.1.

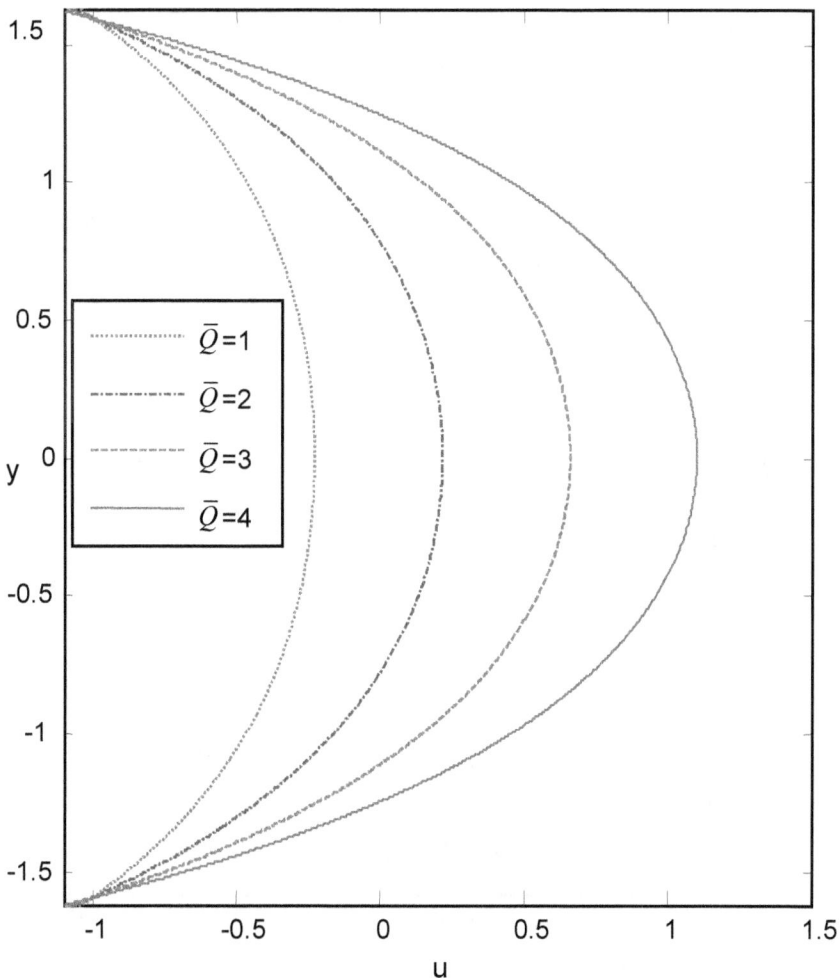

Figure 4.5 The velocity profiles with a=0.5, b=0.5, d=1, x=1, $\phi=\pi/3$, M=1, β =0.1, and Da =0.1.

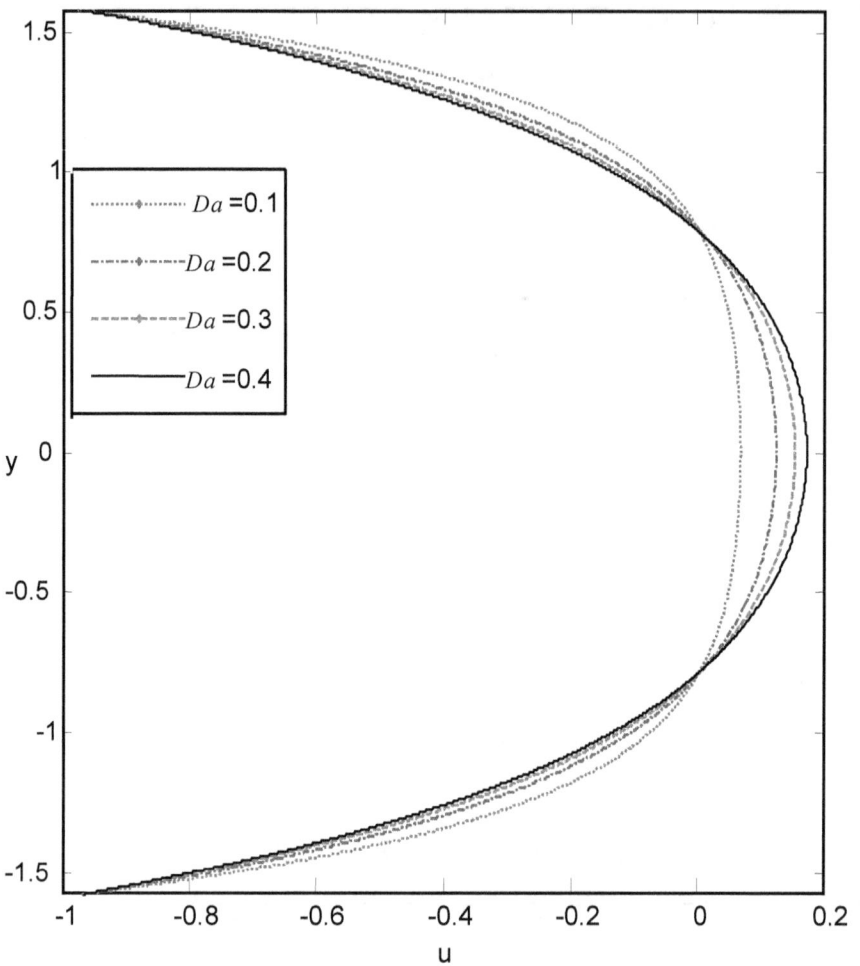

Figure 4.6: The velocity profiles with a=0.5, b=0.5, d=1, x=1, $\phi=\pi/6$, M=1, β =0.1 and $\bar{Q}$=1.

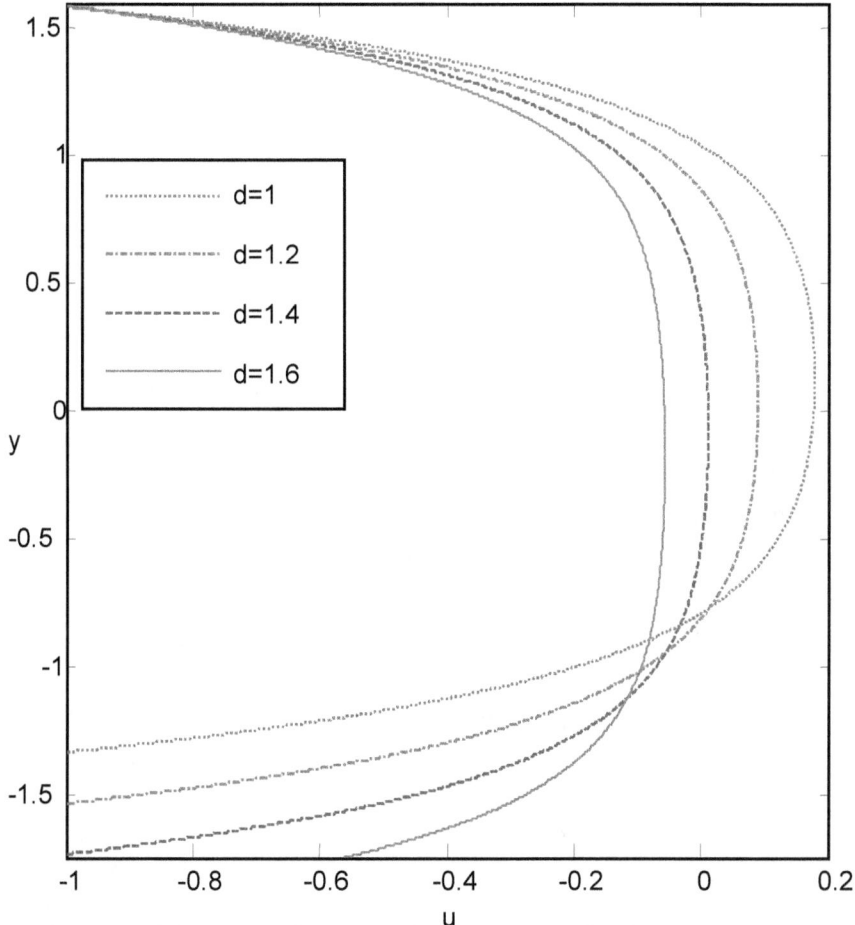

Figure 4.7: The velocity profiles with a=0.5, b=0.5, x=1, $\phi=\pi/6$, M=1,

β =0.1, $\overline{Q}$=2 and Da =0.1.

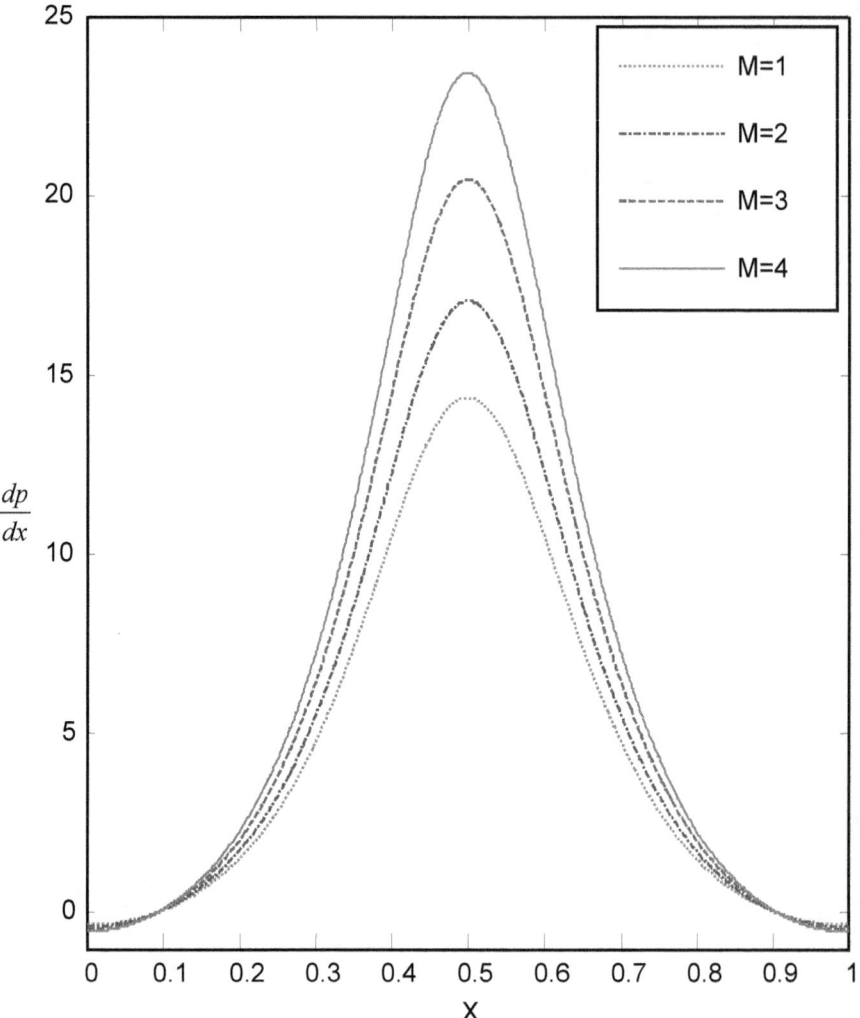

Figure 4.8: Various of pressure gradient $\frac{dp}{dx}$ with x for different values of Hartmann

number M with a=0.6, b=0.6, d=1, ϕ=0, β =0.1, $\overline{Q}$ =1 and Da =0.1.

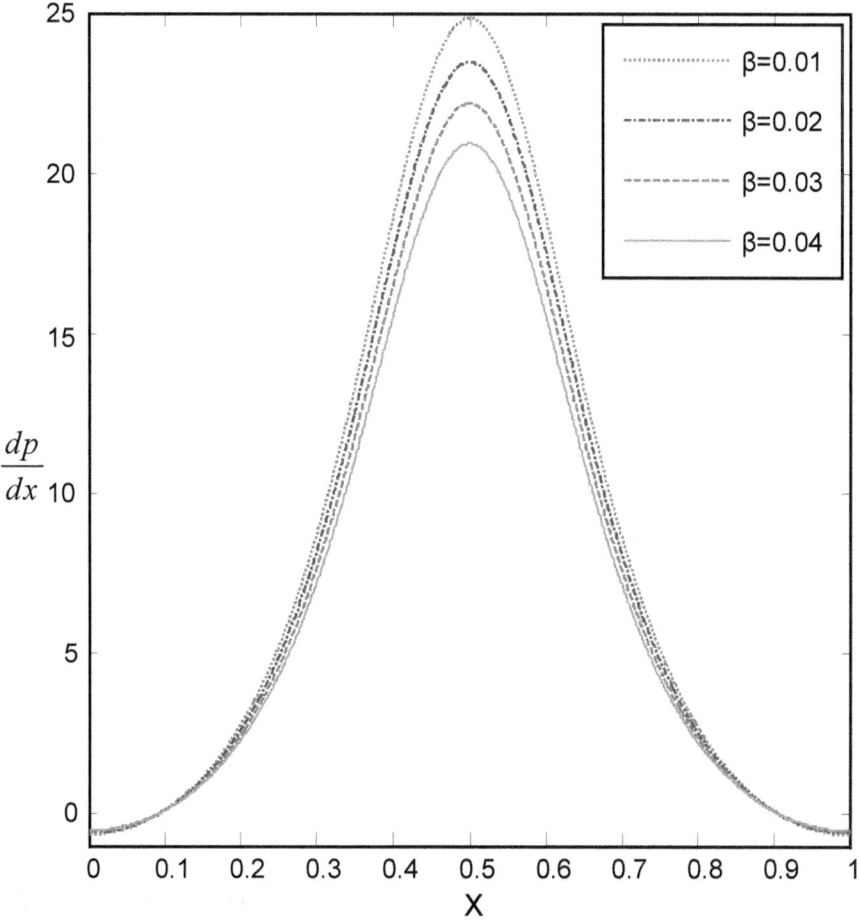

Figure 4.9: Various of pressure gradient $\dfrac{dp}{dx}$ with x foe different values of

permeability parameter β with a=0.6, b=0.6, d=1, ϕ=0, M=1, $\overline{Q}$=1 and Da
=0.1.

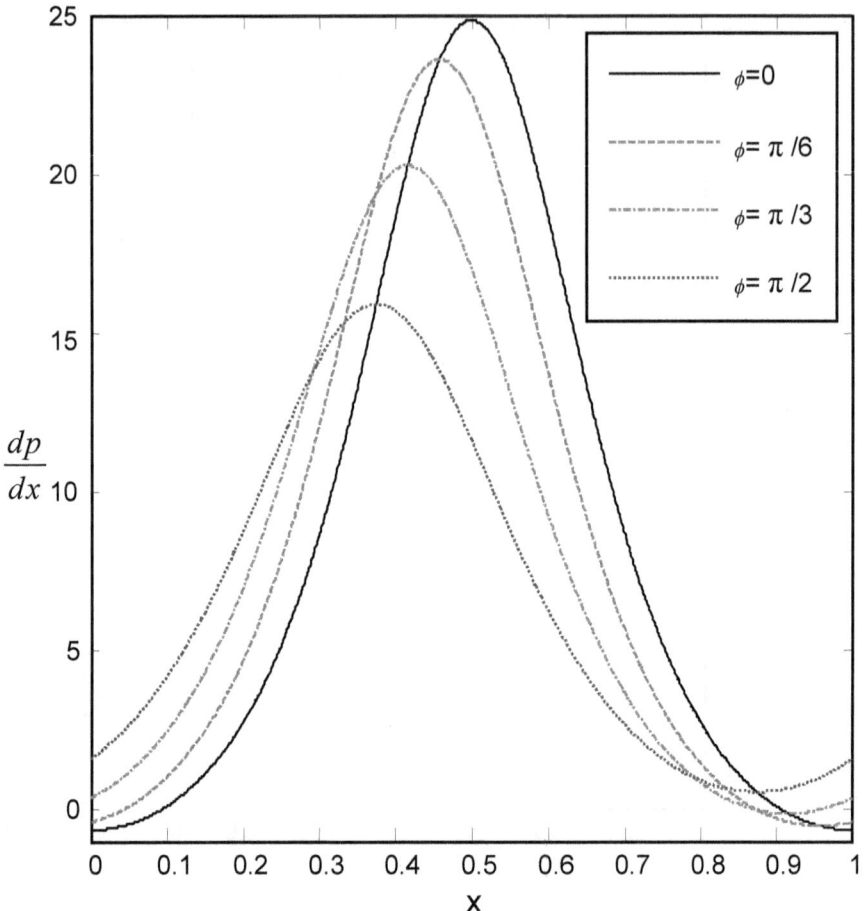

Figure 4.10: Various of pressure gradient $\dfrac{dp}{dx}$ with x for different values of phase difference ϕ with a=0.6, b=0.6, d=1, M=1, β =0.01, $\overline{Q}$=1 and Da =0.1.

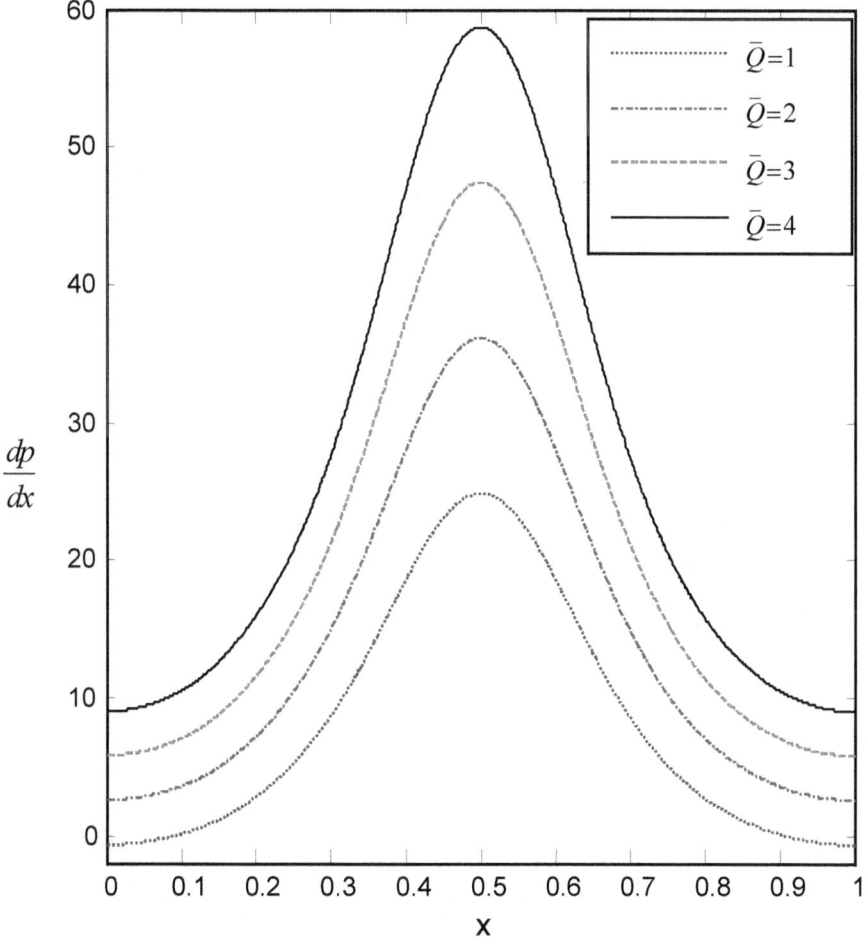

Figure 4.11: Various of pressure gradient $\dfrac{dp}{dx}$ with x for different values of mean

flow rate $\overline{Q}$ with a=0.6, b=0.6, d=1, M=1, ψ=0, β =0.01, and Du =0.1.

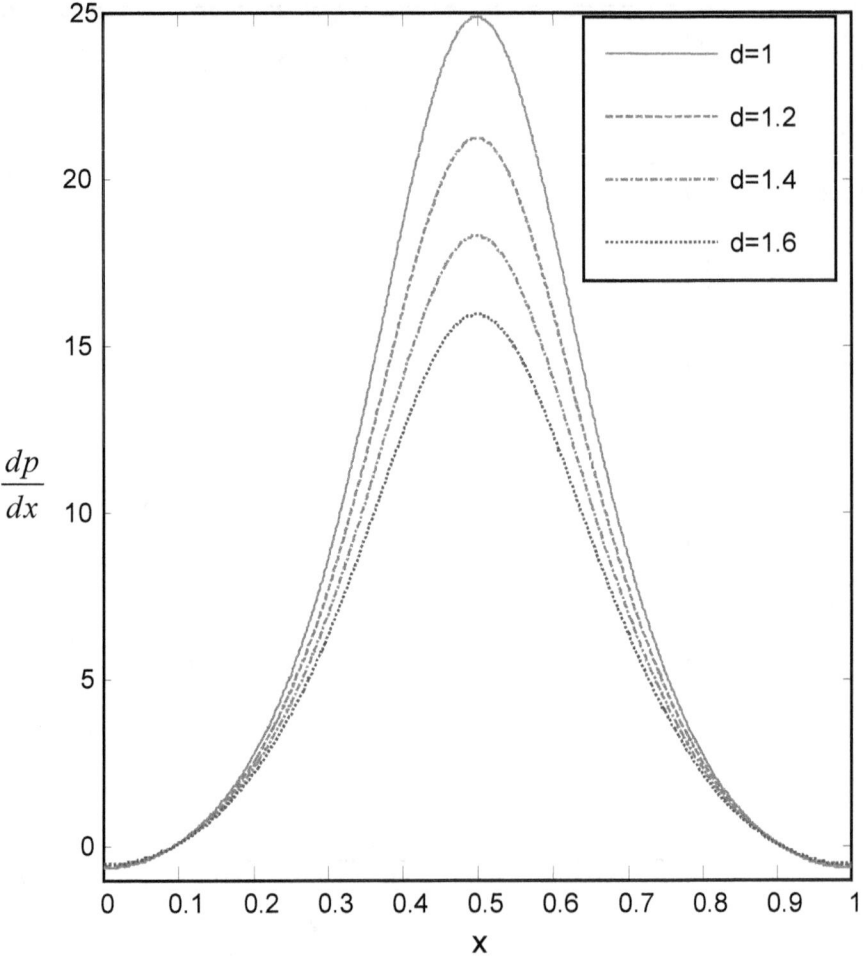

Figure 4.12: Various of pressure gradient $\dfrac{dp}{dx}$ with x for different values of channel

width d with a=0.6, b=0.6, M=1, ϕ=0, β =0.01, $\overline{Q}$=1 and Da =0.1.

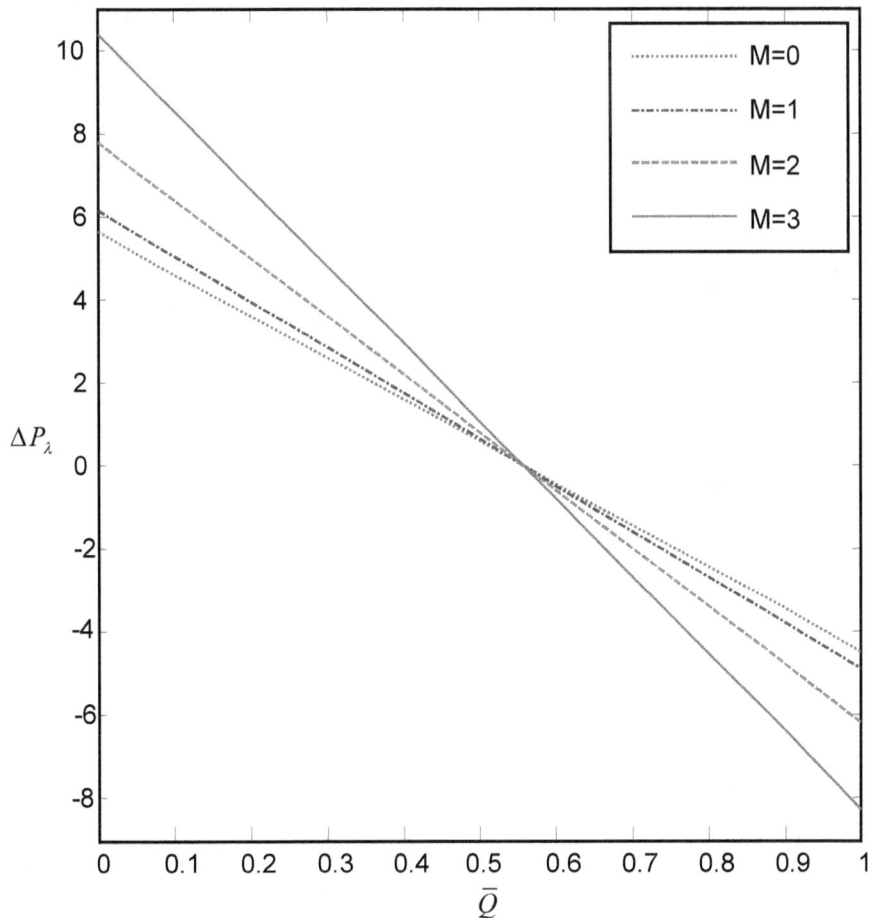

Figure 4.13: Various of $\overline{Q}$ with ΔP_λ for different Hartmann number M with a=0.5, b=0.5, d=1, $\phi=\pi/6$, $\beta=0.01$ and Da =0.1.

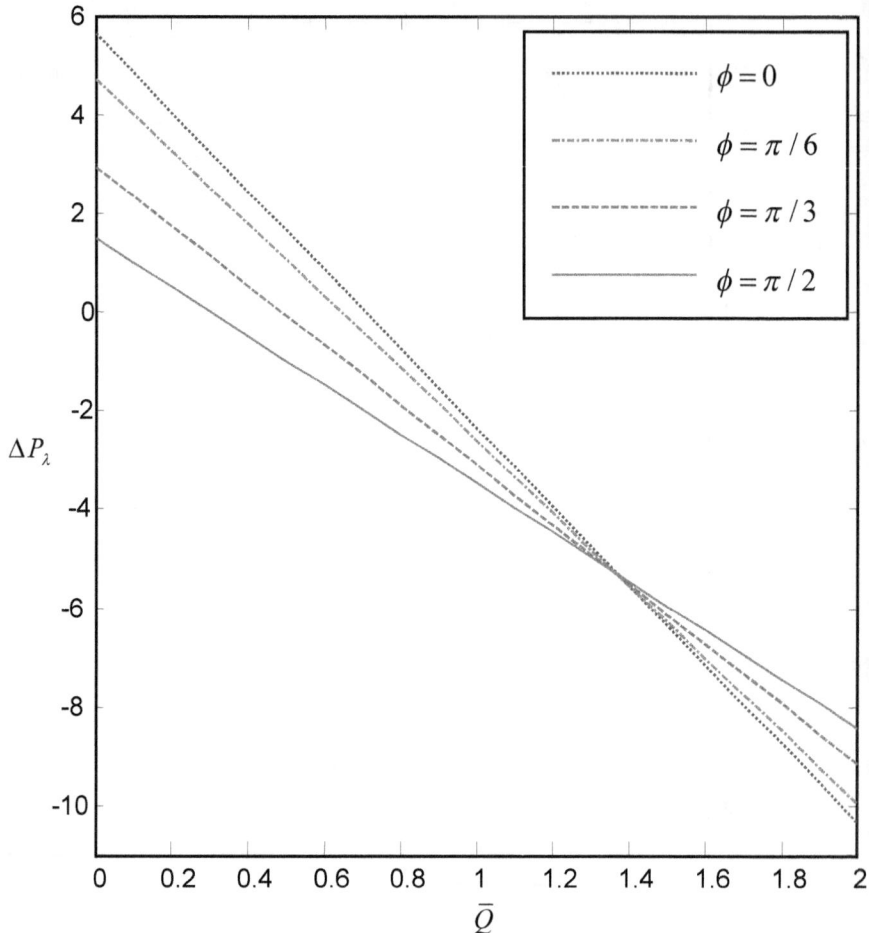

Figure 4.14: Various of $\overline{Q}$ with ΔP_λ for different values of phase difference ϕ with a=0.6, b=0.6, d=0.5, β=0.01, M=1 and Da =0.1.

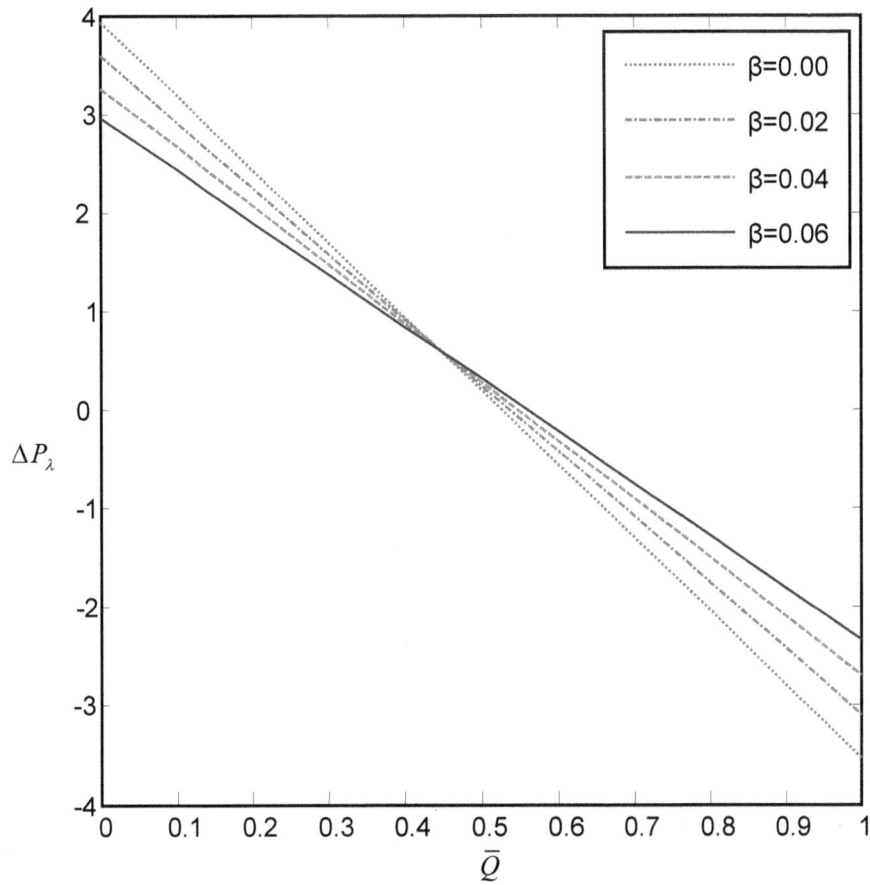

Figure 4.15: Various of $\overline{Q}$ with ΔP_λ for different values of permeability parameter β with a=0.7, b=0.7, d=1, ϕ=π/6, M=1 and Da =0.1.

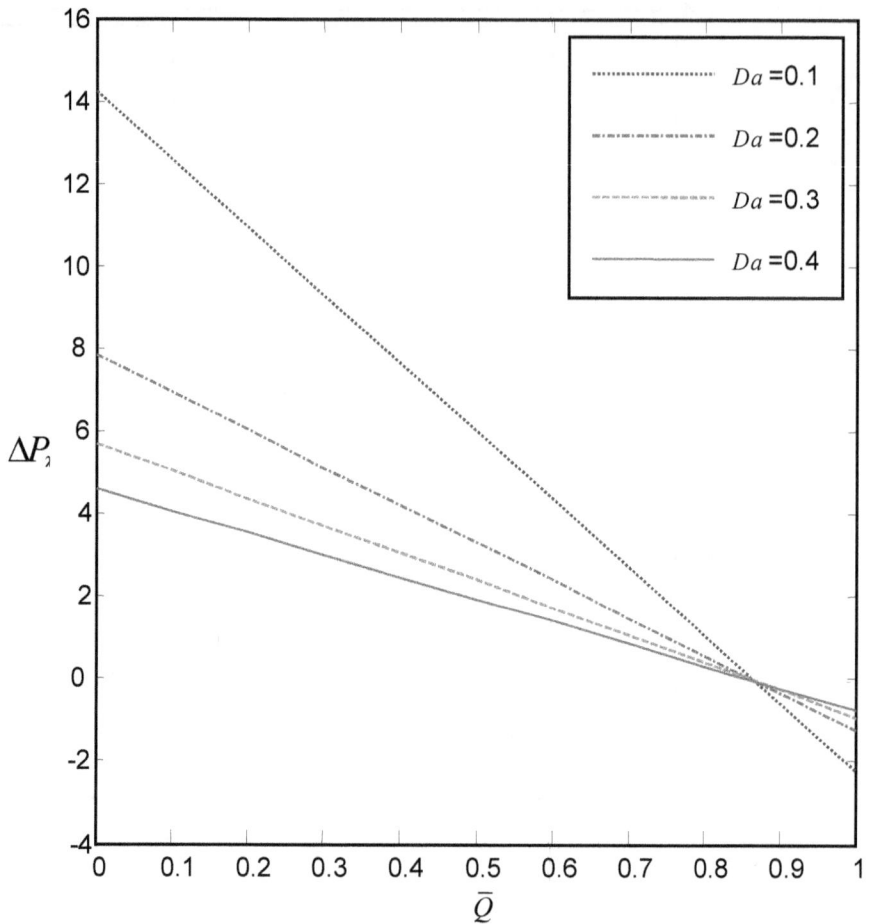

Figure 4.16: Various of $\overline{Q}$ with ΔP_λ for different Darcy number Da with a=0.7, b=0.7, d=1, β=0.01, ϕ=π/6, and M=1.

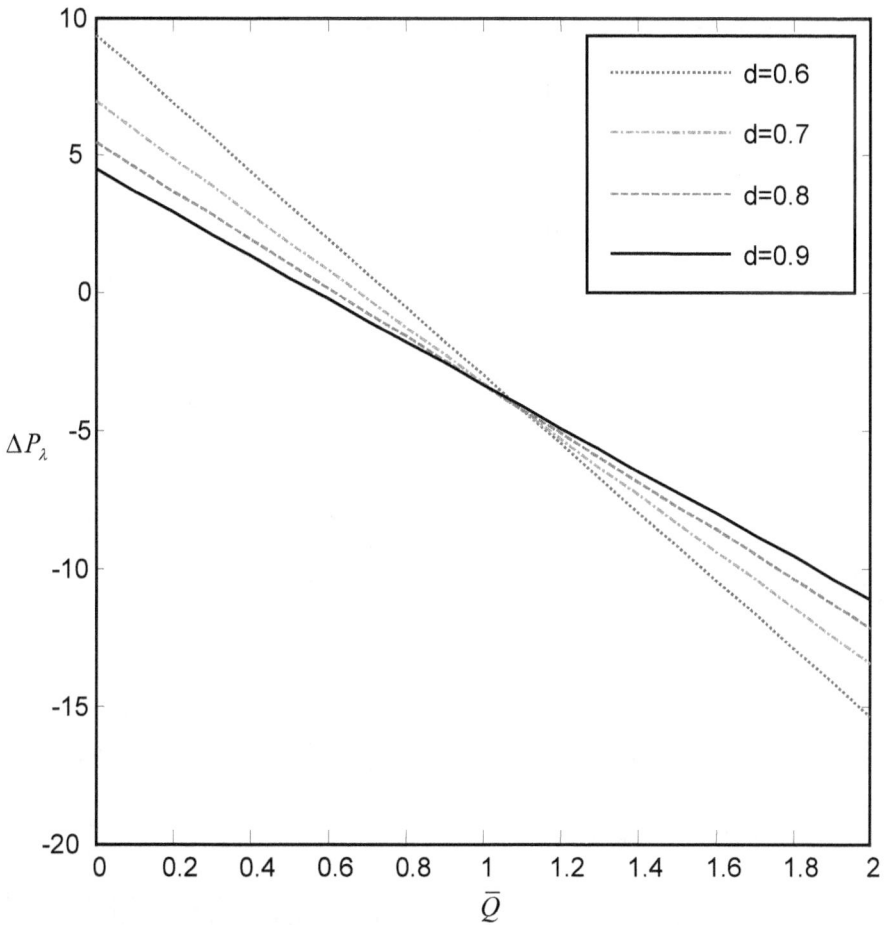

Figure 4.17: Various of $\overline{Q}$ with ΔP_λ for different values of channel width d with

a=0.7, b=0.7, M=1, β=0.01, ϕ=π/6, and Da =0.1

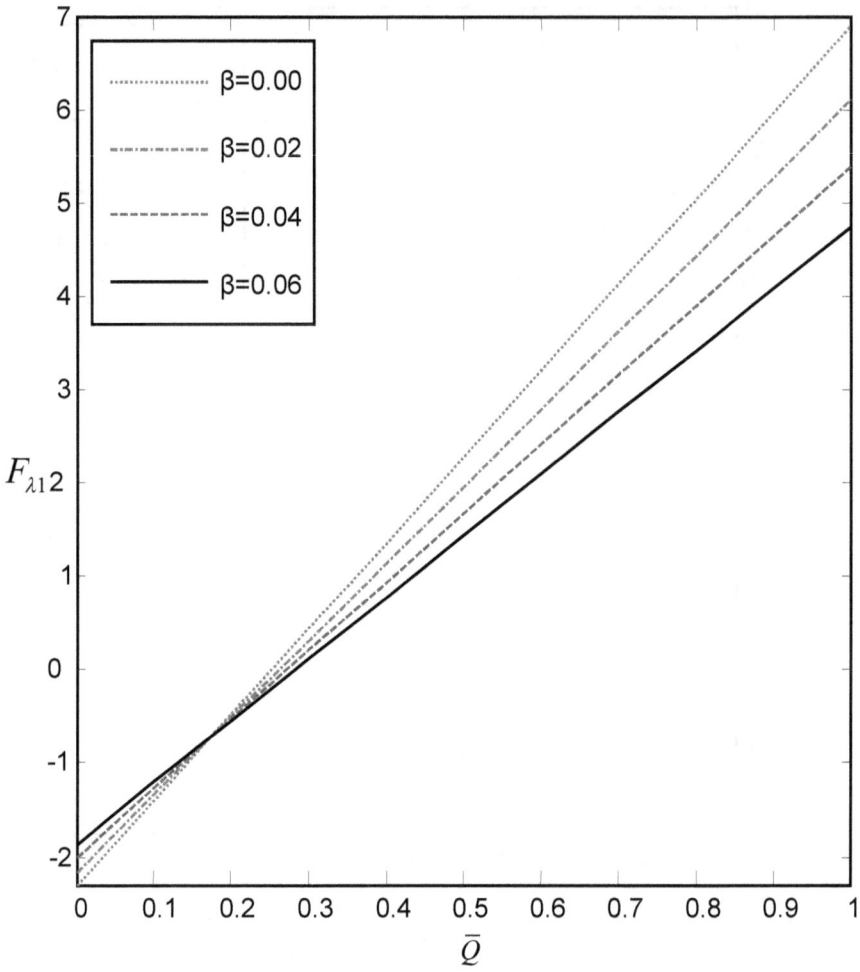

Figure 4.18: Various of $\overline{Q}$ with $F_{\lambda 1}$ for different values of permeability parameter β

with a=0.7, b=0.7, d=0.5, M=1, β=0.01, ϕ=π/6, and Da =0.1

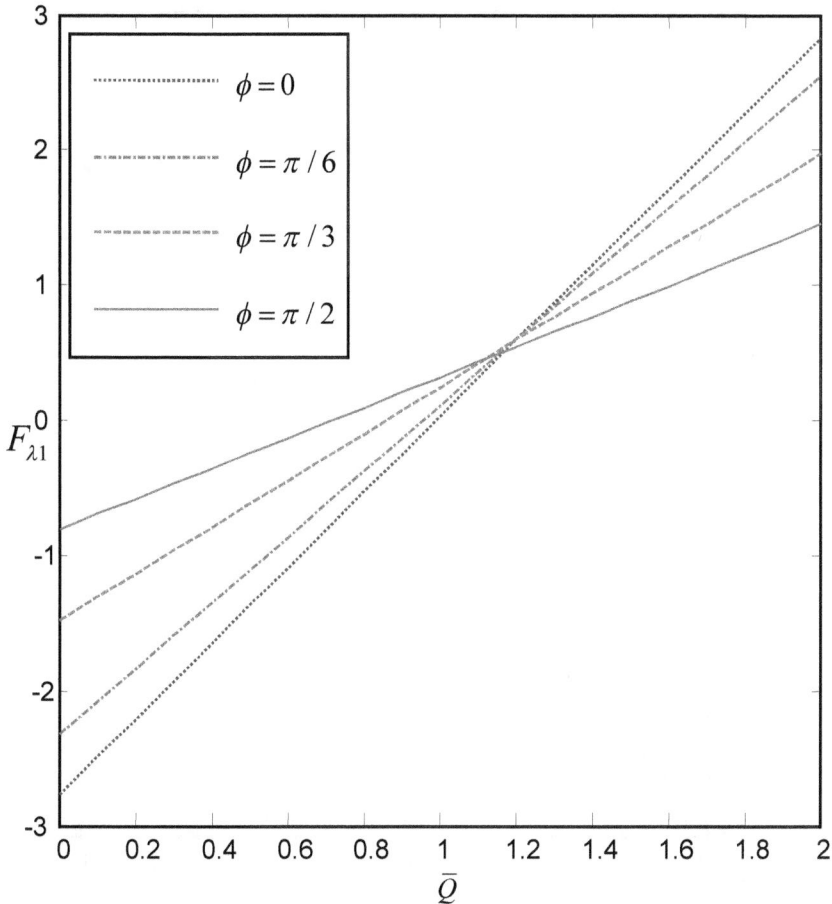

Figure 4.19: Various of $\bar{Q}$ with $F_{\lambda 1}$ for different values of phase difference ϕ with a=0.3, b=0.8, d=0.3, M=1, β=0.3, and Da =0.1.

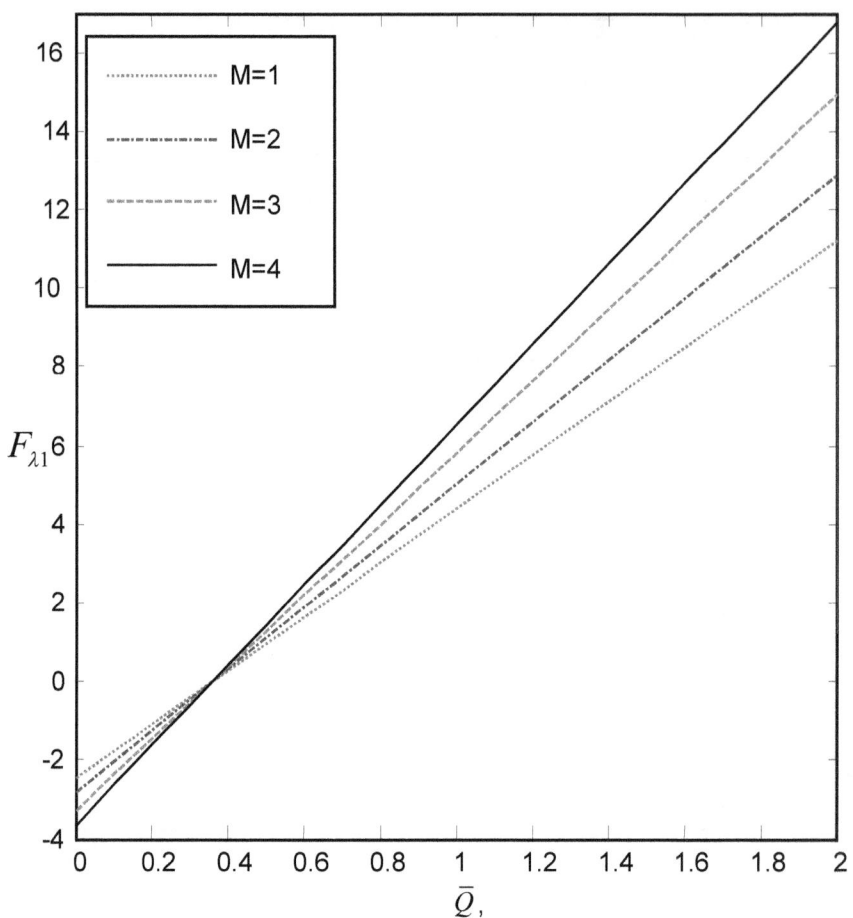

Figure 4.20: Various of $\bar{Q}$ with $F_{\lambda 1}$ for different values of Hartmann number M

with a=0.3, b=0.8, d=0.3, M=1, β=0.1, ϕ=π/3, and Da =0.1

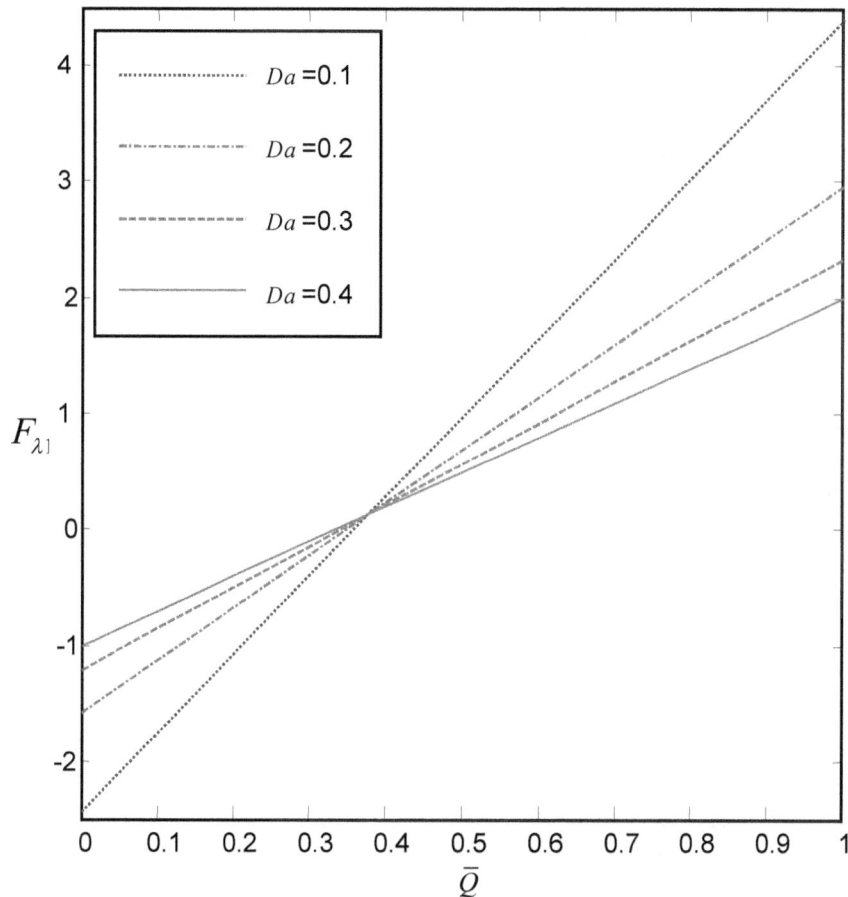

Figure 4.21: Various of $\overline{Q}$ with $F_{\lambda 1}$ for different values of Darcy number Da with a=0.3, b=0.8, d=0.3, M=1, β=0.1, and ϕ=π/3.

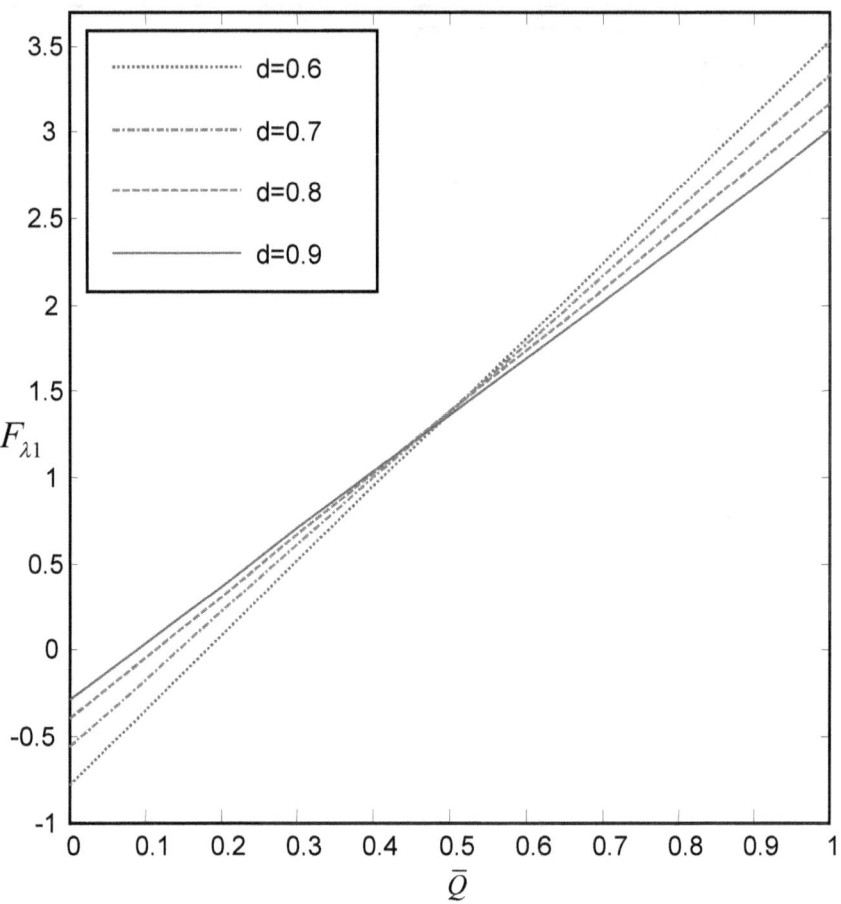

Figure 4.22: Various of $\overline{Q}$ with $F_{\lambda 1}$ for different values of channel width d with

a=0.3, b=0.8, M=1, β=0.1, ϕ=π/3, and Da =0.1

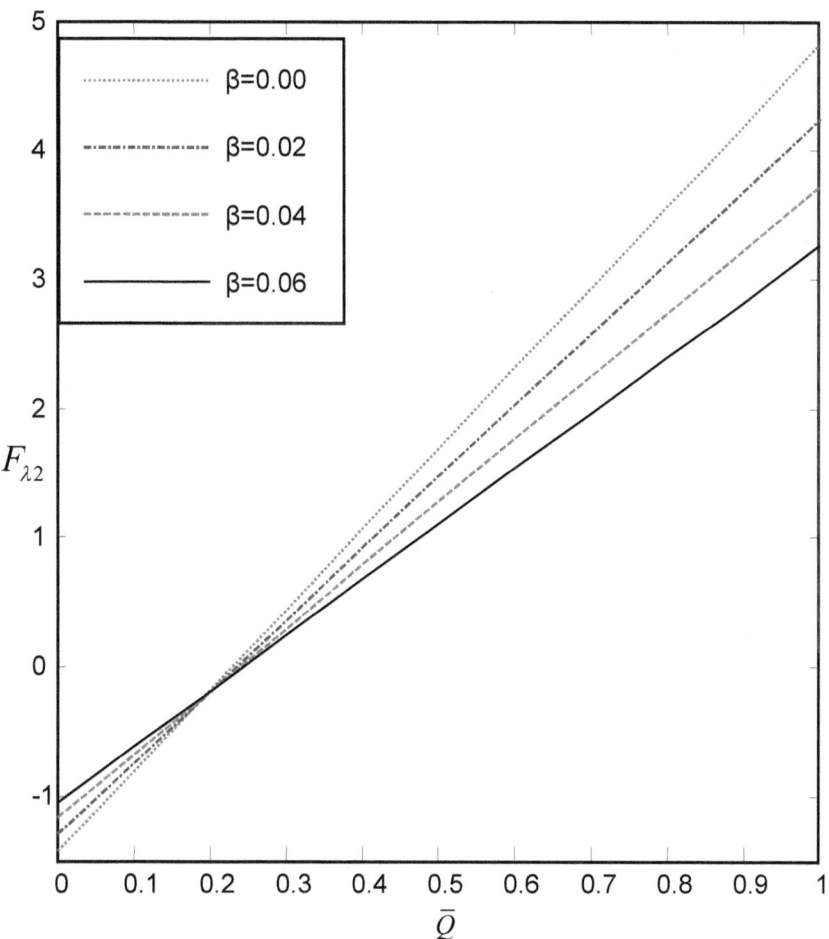

Figure 4.23: Various of $\overline{Q}$ with $F_{\lambda2}$ for different values of permeability parameter β with a=1, b=0.1, d=1, M=1, β=0.1, ϕ=π/6, and Da =0.1.

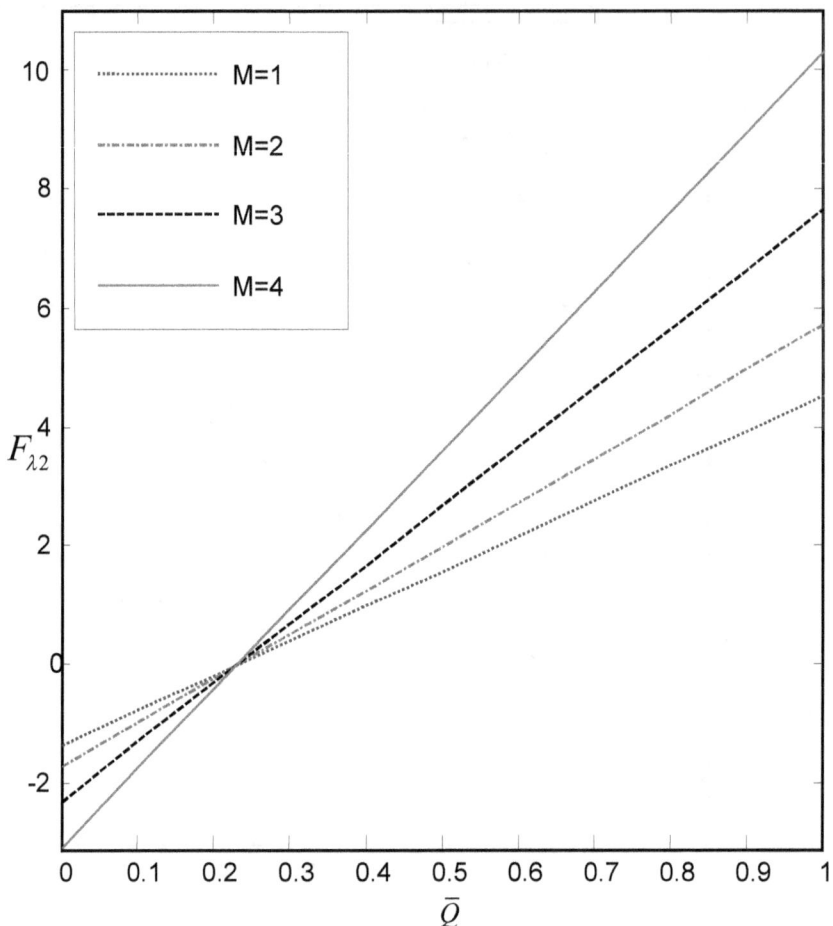

Figure 4.24: Various of $\overline{Q}$ with F_{λ_2} for different values of Hartmann number M with a=1, b=0.1, d=1, M=1, β=0.1, ϕ=π/6, and Da =0.1.

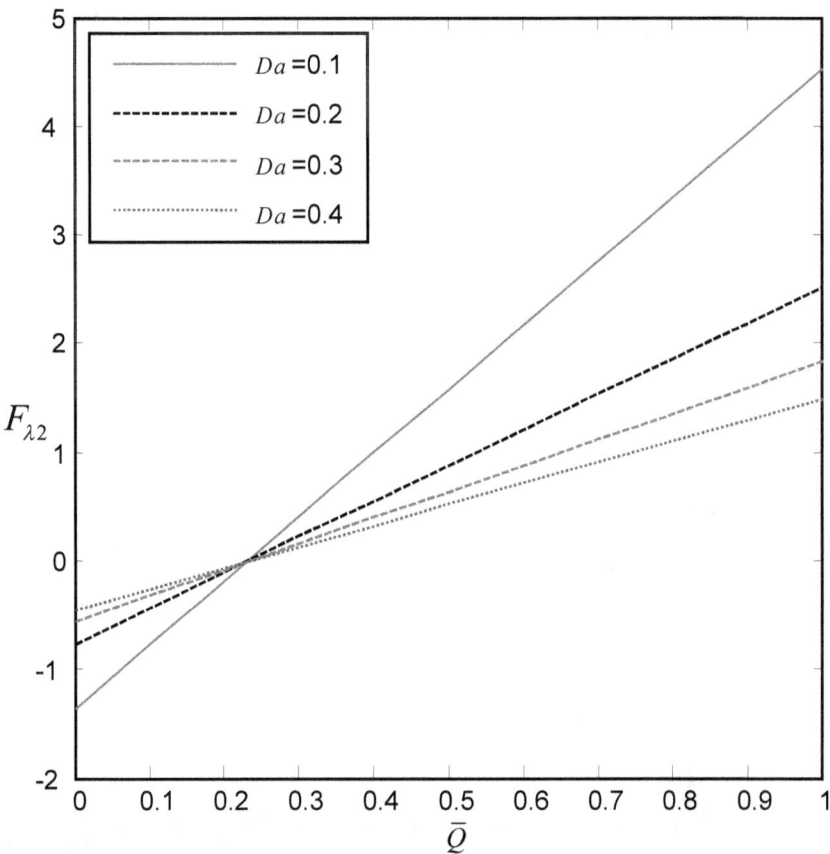

Figure 4.25: Various of $\overline{Q}$ with $F_{\lambda 2}$ for different values of Darcy number Da with a=1, b=0.1, d=1, M=1, β=0.1 and ϕ=π/6.

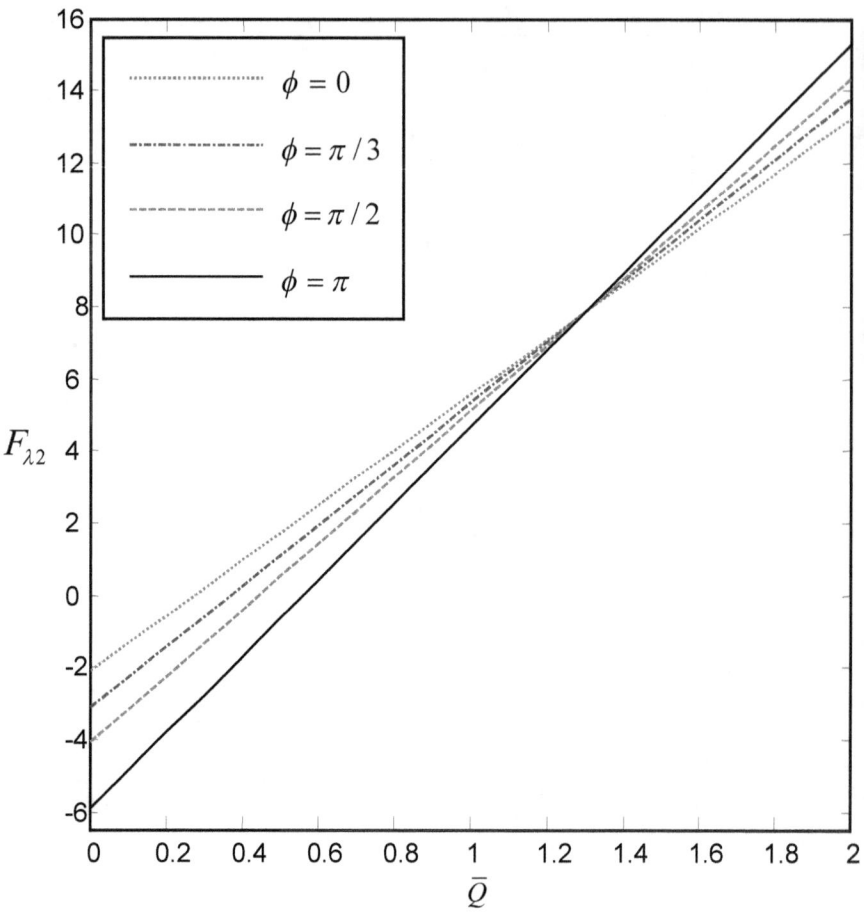

Figure 4.26: Various of $\overline{Q}$ with $F_{\lambda 2}$ for different values of phase difference ϕ with a=1, b=0.1, d=1, M=1, β=0.01 and Da =0.01.

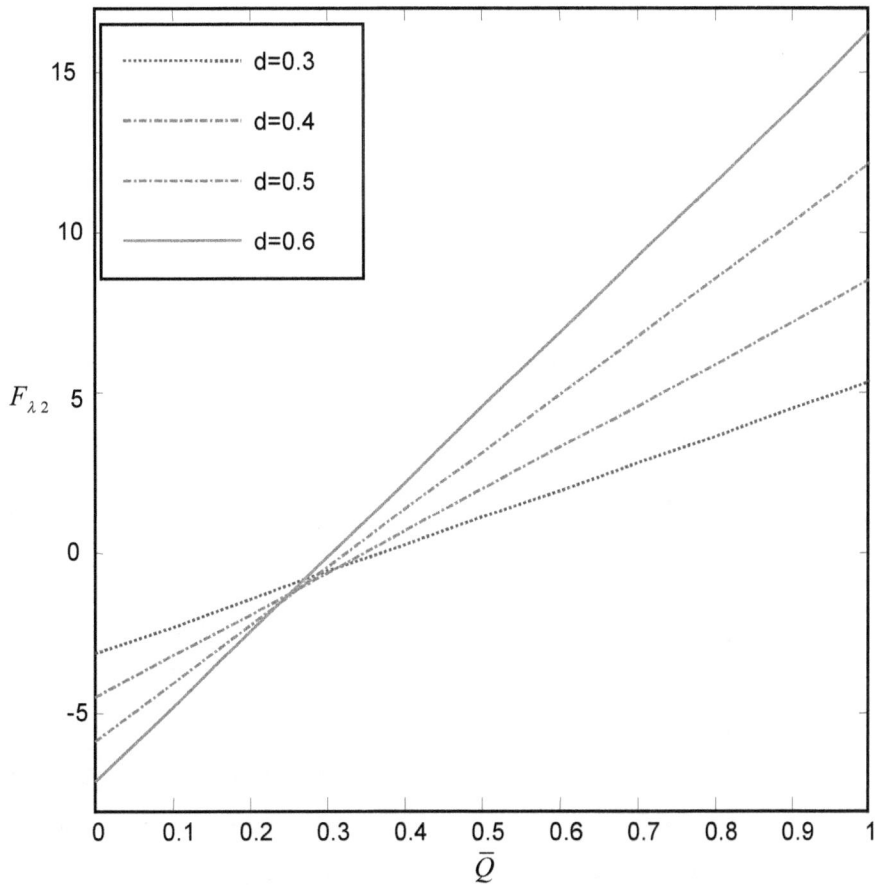

Figure 4.27: Various of $\overline{Q}$ with $F_{\lambda 2}$ for different values of channel width d with

a=1, b=0.1, d=1, M=1, β=0.01, ϕ=π/3 and Da =0.01.

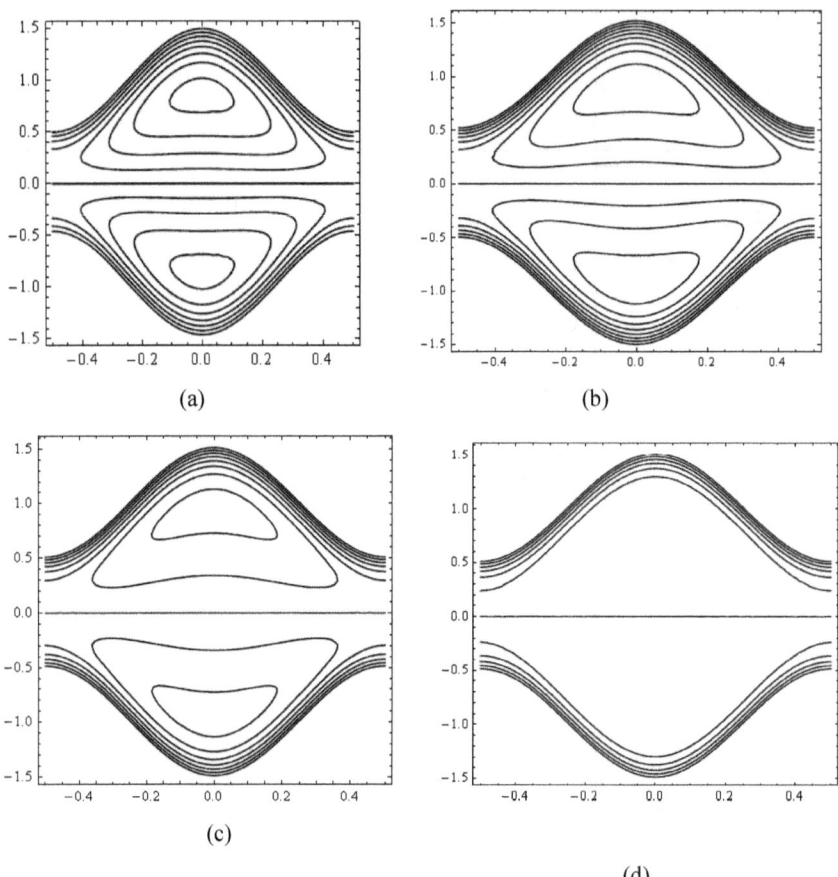

(a)

(b)

(c)

(d)

Figure 4.28: Stream lines for different values of Hartmann number M and fixed parameters chosen as a=0.5, b=0.5, d=1, ϕ=0, β=0.01, $\overline{Q}$=1.8 and Da =0.1,(a) M=0, (b) M=1, (c) M=2 and (d) M=3. □

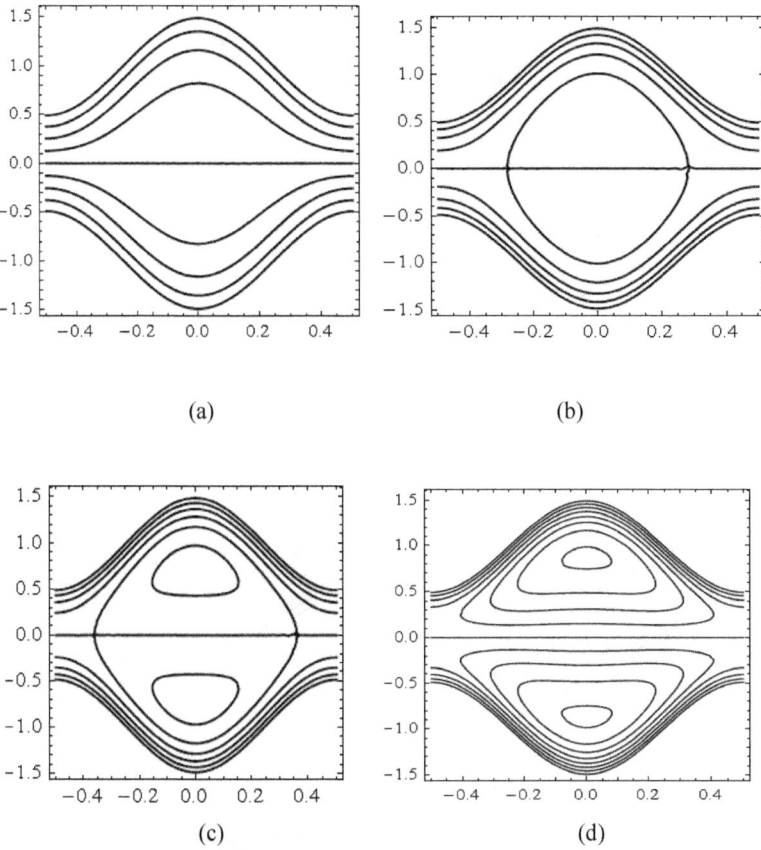

Figure 4.29 Stream lines for different values of the mean flow rate $\bar{Q}$ and fixed
parameters chosen as a=0.5, b=0.5, d=1, ϕ=0, β=0.01 M=1 and Da=0.2, (a)
$\bar{Q}$=1.2, (b) $\bar{Q}$=1.4, (c) $\bar{Q}$=1.6 and (d) $\bar{Q}$=1.8.

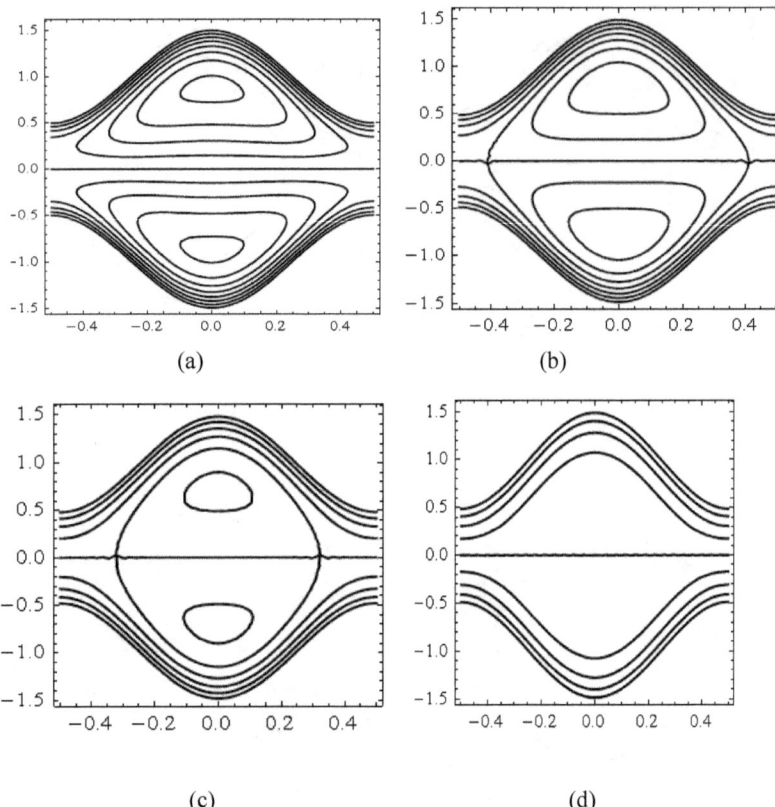

(a)   (b)

(c)   (d)

Figure 4.30: Stream lines for different values of the permeability parameter β and fixed parameters chosen as a=0.5, b=0.5, d=1, ϕ=0, $\overline{Q}$=1.8, M=1 and Da =0.2. (a) β =0, (b) β =0.2, (c) β =0.4 and (d) β =0.6.

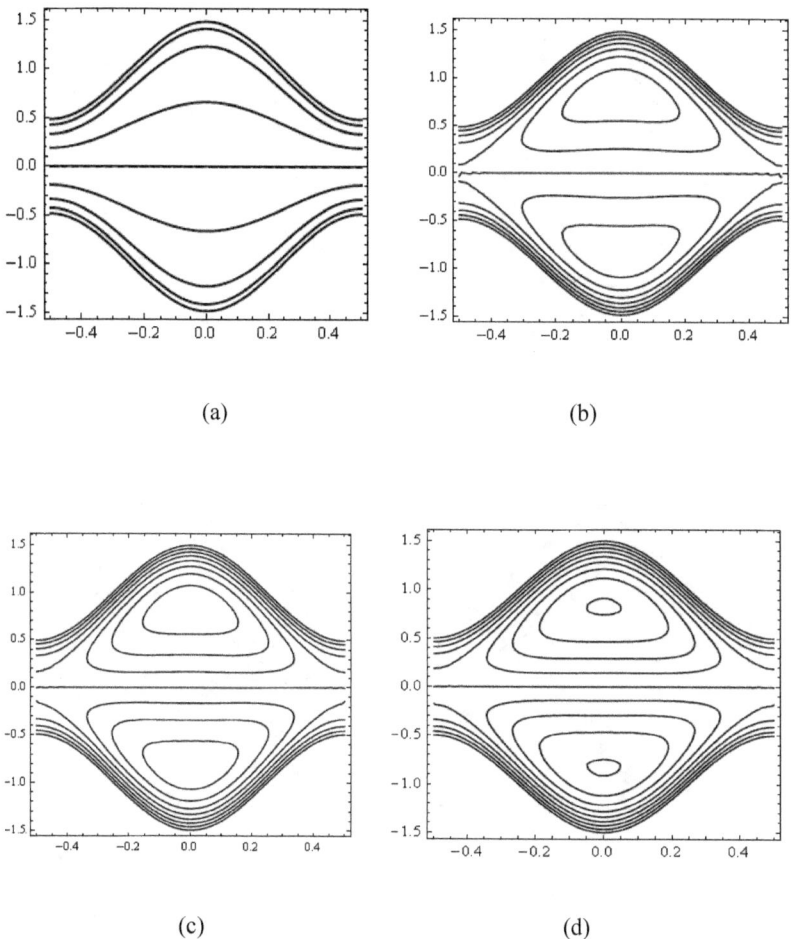

Figure 4.31: Stream lines for different values of the Darcy number Da and   fixed

parameters chosen as a=0.5, b=0.5, d=1, ϕ=0, $\overline{Q}$=1.8, M=1 and

β =0.1. (a) Da =0.01, (b) Da =0.1, (c) Da =0.2 and (d) Da =0.3.□

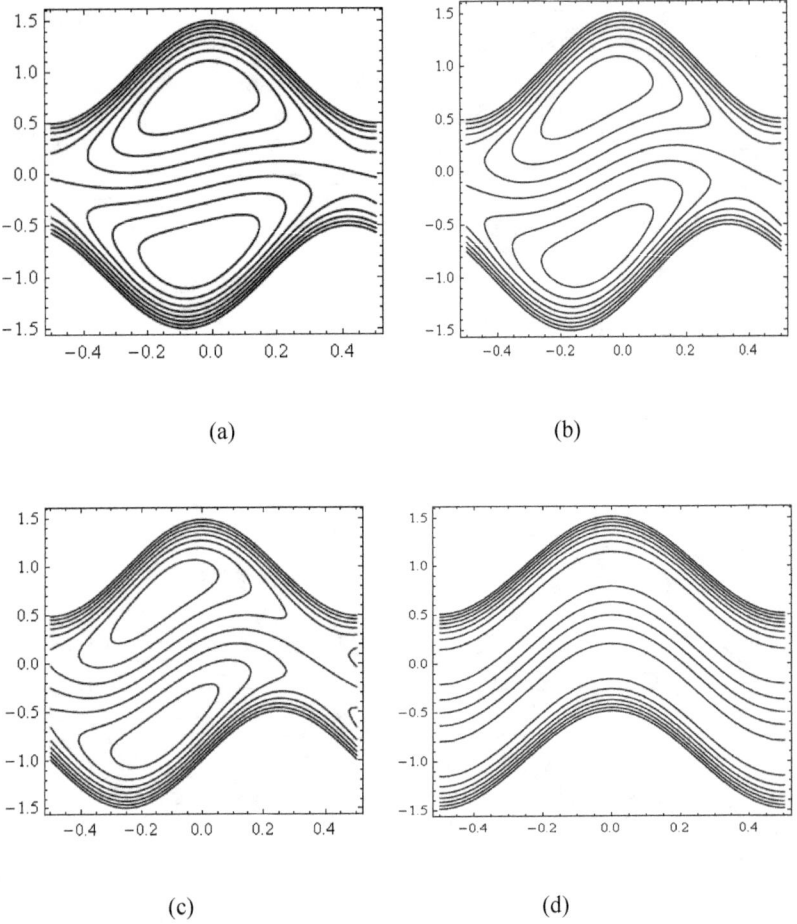

(a)

(b)

(c)

(d)

Figure 4.32: Stream lines for different values of the phase difference Φ and fixed parameters chosen as a=0.5, b=0.5, d=1, $\overline{Q}$=1.8, M=1, β =0.1 and Da =0.3, (a) ϕ=π/6, (b) ϕ= π/3, (c) ϕ= π/2 and (d) ϕ= π. □

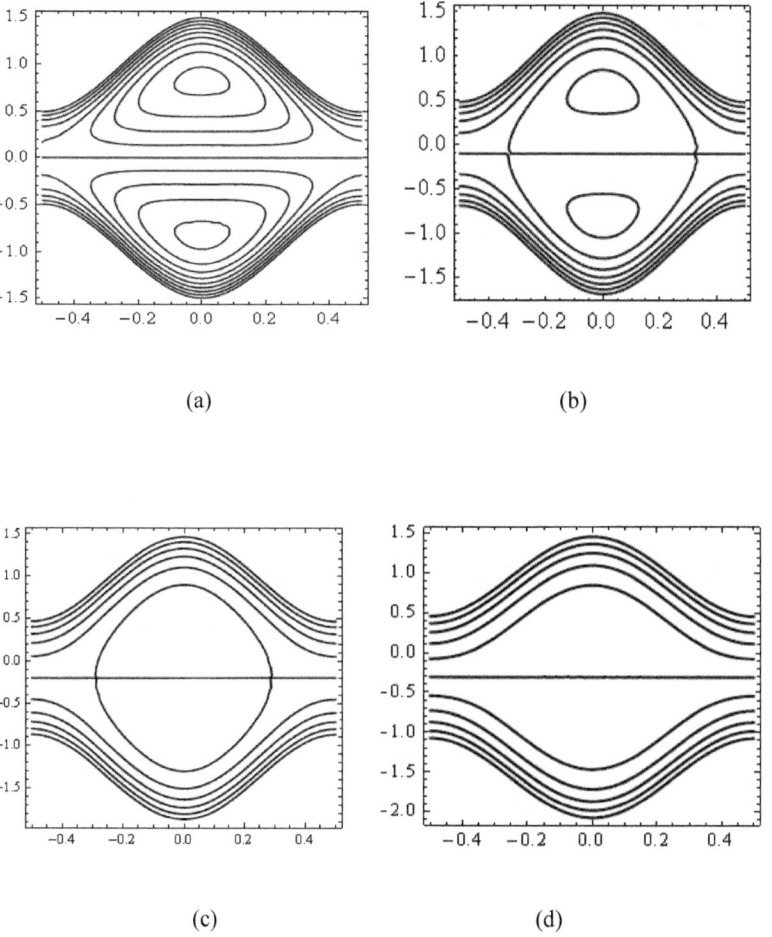

(a)   (b)

(c)   (d)

Figure 4.33: Stream lines for different values of the channel width d and fixed parameters chosen as a=0.5, b=0.5, $\overline{Q}$=1.8, M=1, ϕ=0, β =0.1 and Da =0.3, (a) d=1, (b) d=1.2, (c) d=1.4 and (d) d=1.6.

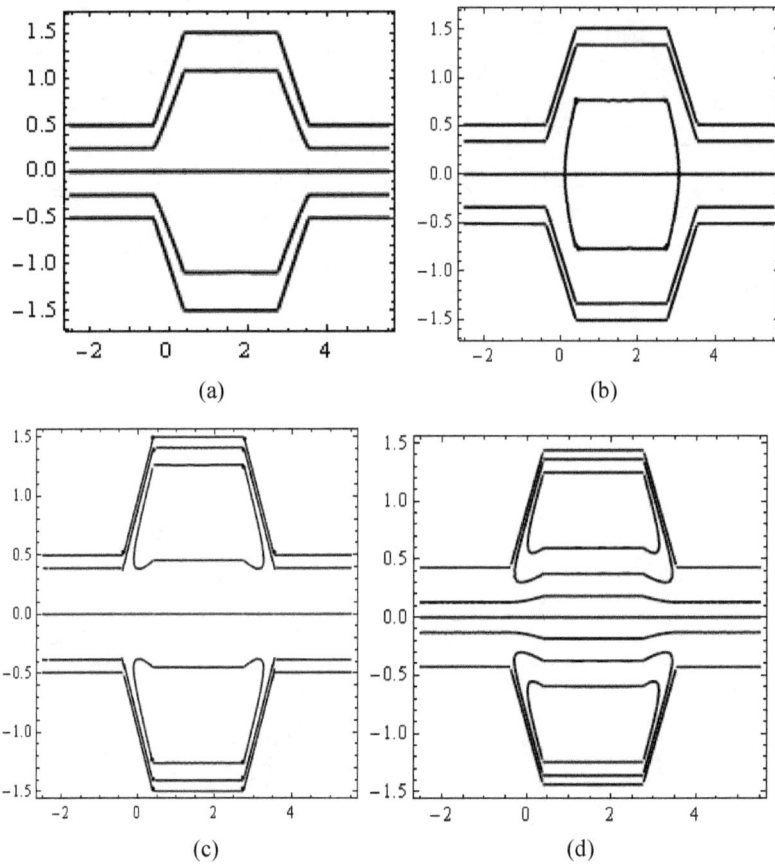

Figure 4.34: Trapezoidal wave of stream lines for different values of the average volume flow rate $\overline{Q}$ and fixed parameters chosen as a=0.5, b=0.5, d=1, M=1, ϕ=0, β =0.01 and Da =0.1, (a) $\overline{Q}$=1, (b) $\overline{Q}$=1.5, (c) $\overline{Q}$=1.8 and (d) $\overline{Q}$=2.

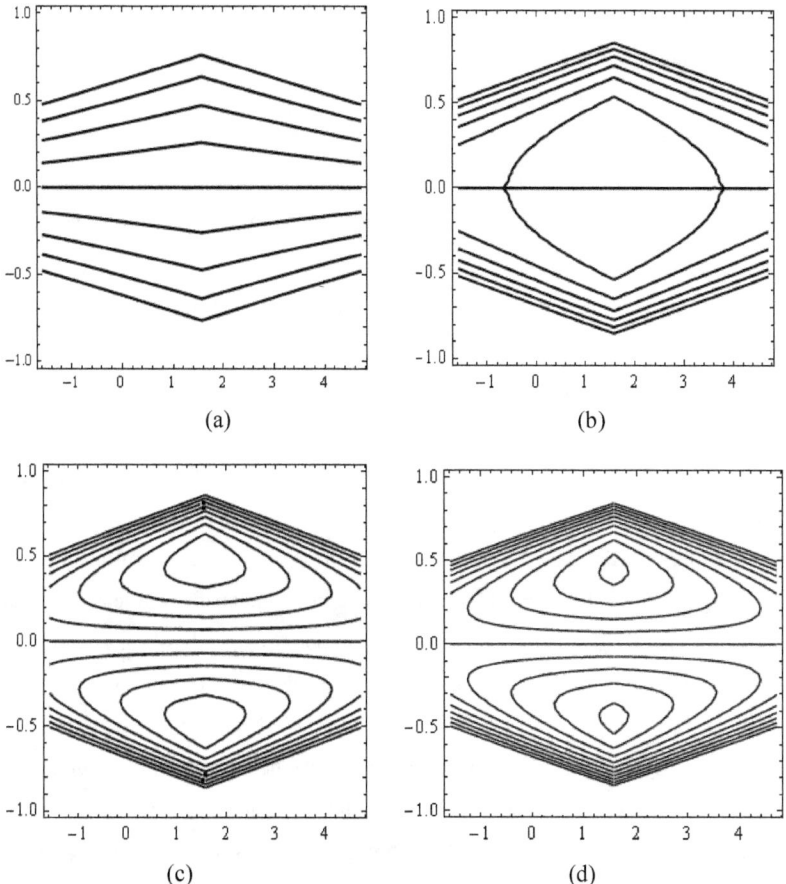

Figure 4.35: Triangular wave of stream lines for different values of the average volume flow rate $\overline{Q}$ and fixed parameters chosen as a=0.5, b=0.5, d=1, M=1, ϕ=0, β =0.01 and Da =0.1, (a) $\overline{Q}$ =1, (b) $\overline{Q}$ =1.5, (c) $\overline{Q}$ =1.8 and (d) $\overline{Q}$ =2.

197

5.1 Introduction

Peristaltic motion in a channel/tube is now known as an important type of flow occurring in several engineering and physiological processes. In particular such flows occur in roller and finger pumps, chyme motion in the gastrointestinal tract, movement of ovum in the fallopian tube, powder technology and many others. Extensive literature on the topic is now available when the fluid occupies the non-porous space. Mekheimer (2003) studied nonlinear peristaltic transport through a porous medium in an inclined planar channel and extended the same analysis to nonlinear peristaltic transport of MHD flow through a porous medium. Hayat *et al.* (2007) observed hall effects on peristaltic flow of Maxwell fluid in a porous medium. Kothandapani and Srinivas (2008) discussed the peristaltic transport in an asymmetric channel with heat transfer. Hayat (2009) analysed the effect of the heat transfer on the peristaltic flow of an electrically conducting fluid in a porous space.

So far, no investigation is made to the peristaltic flow of electrically conducting viscous fluid filling the porous space in an inclined asymmetric channel with heat transfer. Such consideration is very important since bio-heat is currently considered as heat transfer in the human body. In view of this thermotherapy and the human thermoregulation system, the model of bio-heat transfer in tissues has been attracted by the biomedical engineers. In fact the heat transfer in human tissues involves complicated processes such as heat conduction in tissues, heat transfer due to perfusion of the arterial-venous blood through the pores of the tissue, metabolic heat generation and external interactions such as electromagnetic radiation emitted from cell phones. It is observed that the all the biological organs in a human body are not horizontal and hence these organs are to be modeled as inclined tubes/channels. Furthermore, the electrically conducting effect on peristaltic flow in an inclined asymmetric channel in a porous space is important in technology (for example, MHD pump) and biology (for example, blood flow). Such analysis is of great value in medical research. In view of these facts the aim of present investigation is to extend the work of in two directions. Firstly to consider a fluid filling the porous space in an inclined asymmetric channel and secondly to analyze the heat transfer effects. The flow modeling is based upon continuity, momentum and energy equations. The compatibility and energy equations are first expressed in terms of the stream function and then solved in closed form for long wavelength and low Reynolds number assumptions. The obtained expressions are also utilized to discuss the influence of emerging parameters on the flow quantities. In view of these facts, the influence of porous space on peristaltic transport of electrically conducting fluid is investigated. The velocity and the stream function are determined. The pressure rise per one wave length is calculated and the effects of various physical parameters on the pumping characteristics are analyzed.

5.2 Mathematical Formation

Consider an incompressible viscous fluid filling porous space in an inclined asymmetric channel the fluid is an electrically conducting in the presence of an applied magnetic field $\bar{B}_0$. The induced magnetic field is neglected. The temperature of the upper and lower walls are T_0 and T_1, respectively (Figure 5.1). The sinusoidal waves propagating along the channel wall are of the following forms:

$$h_1\left(\bar{X},\bar{t}\right)=d_1+a_1\cos\left[\frac{2\pi}{\lambda}\left(\bar{X}-c\bar{t}\right)\right]\ldots\ldots\text{upper wall,}\qquad(5.1)$$

$$h_2\left(\bar{X},\bar{t}\right)=-d_2-b_1\cos\left[\frac{2\pi}{\lambda}\left(\bar{X}-c\bar{t}\right)+\phi\right]\ldots\ldots\text{lower wall.}\qquad(5.2)$$

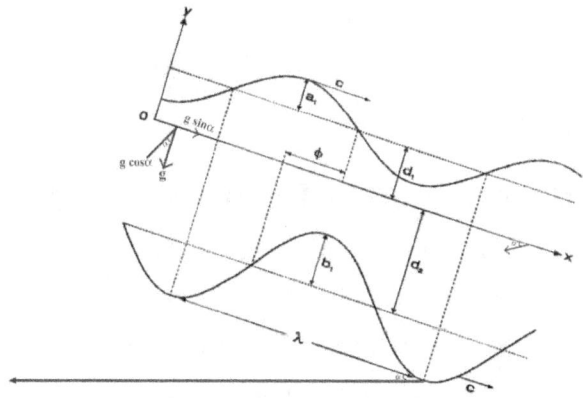

Figure 5.1. Schematic diagram of a two-dimensional asymmetric channel

199

in which a_1 and b_1 are the amplitudes of the waves, λ is the wave length, c is the wave speed, ϕ $(0 \le \phi \le \pi)$ is the phase difference, $\overline{X}$ measured along the axis of the channel and $\overline{Y}$ perpendicular to $\overline{X}$. Let $(\overline{U}, \overline{V})$ be the velocity components in fixed frame of reference $(\overline{X}, \overline{Y})$. It should be noted that $\phi = 0$ corresponding to symmetric channel with waves out of phase and for $\phi = \pi$ the waves are in phase, and further a_1, b_1, d_1, d_2 and ϕ satisfies the condition

$$a_1^2 + b_1^2 + 2a_1b_1 \cos\phi \le \left(d_1 + d_2\right)^2 .\tag{5.3}$$

5.3 Equations of motion

The corresponding flow equations in the porous medium are

$$\frac{\partial \overline{U}}{\partial \overline{X}} + \frac{\partial \overline{V}}{\partial \overline{Y}} = 0,\tag{5.4}$$

$$\frac{\partial \overline{U}}{\partial \overline{X}} + \overline{U}\frac{\partial \overline{U}}{\partial \overline{X}} + \overline{V}\frac{\partial \overline{U}}{\partial \overline{Y}} = -\frac{1}{\rho}\frac{\partial \overline{P}}{\partial \overline{X}} + v\left(\frac{\partial^2 \overline{U}}{\partial \overline{X}^2} + \frac{\partial^2 \overline{U}}{\partial \overline{Y}^2}\right) - \frac{v}{K}\overline{U} - \sigma B_0^2 \overline{U} + \rho g \sin\alpha ,\tag{5.5}$$

$$\frac{\partial \overline{V}}{\partial \overline{t}} + \overline{U}\frac{\partial \overline{V}}{\partial \overline{X}} + \overline{V}\frac{\partial \overline{V}}{\partial \overline{Y}} = -\frac{1}{\rho}\frac{\partial \overline{P}}{\partial \overline{X}} + v\left(\frac{\partial^2 \overline{V}}{\partial \overline{X}^2} + \frac{\partial^2 \overline{V}}{\partial \overline{Y}^2}\right) - \frac{v}{K}\overline{V} - \rho g \cos\alpha ,\tag{5.6}$$

$$\xi\left[\frac{\partial}{\partial t} + \overline{U}\frac{\partial}{\partial \overline{X}} + \overline{V}\frac{\partial}{\partial \overline{Y}}\right]T = \frac{k}{\rho}\Delta^2 T + v\left[2\left\{\left(\frac{\partial \overline{U}}{\partial \overline{X}}\right)^2 + \left(\frac{\partial \overline{U}}{\partial \overline{Y}}\right)^2\right\} + \left(\frac{\partial \overline{U}}{\partial \overline{Y}} + \frac{\partial \overline{V}}{\partial \overline{X}}\right)^2\right].\tag{5.7}$$

where

$$\Delta^2 = \frac{\partial^2}{\partial X^2} + \frac{\partial}{\partial Y^2} .$$

In the above equations $\overline{U}$ and $\overline{V}$ are the respective velocity components in the $\overline{X}$ and $\overline{Y}$ directions of the fixed frame, ρ is the constant density of the fluid, $\overline{P}$ is the fluid pressure, μ and v are the respectively dynamic and kinematic viscosity, $\overline{K}$ is permeability of the porous medium, k is the thermal conductivity, ξ is the specific heat at constant volume and T is the temperature. Bars indicate that the respective quantity is inherently unsteady in the fixed frame but it can be treated as steady by switching form the fixed frame to the wave frame. If $\bar{u}$ and $\bar{V}$ denote the velocity components in the $\bar{x}$ and $\bar{y}$ directions of the wave frame, then

$$\bar{x} = \bar{X} - ct, \bar{y} = \bar{Y}, \bar{u} = \bar{U} - c, \bar{v} = \bar{V}, \bar{p}(x) = \bar{P}(\bar{X}, \bar{t}). \qquad (5.8)$$

where c is the speed of the peristaltic wave.

We define the following non-dimensional quantities:

$$x = \frac{\bar{X}}{\lambda}, \quad y = \frac{\bar{y}}{d_1}, \quad u = \frac{\bar{u}}{c}, \quad v = \frac{\bar{v}}{c\delta}, \quad \delta = \frac{d_1}{\lambda}, \quad d = \frac{d_2}{d_1}, \quad p = \frac{d_1^2 \bar{p}}{\mu c \lambda}, \quad t = \frac{c\bar{t}}{\lambda}, \quad h_1 = \frac{H_1}{d_1},$$

$$h_2 = \frac{H_2}{d_1}, \quad a = \frac{a_1}{a_1}, \quad b = \frac{b_1}{d_1}, \quad \mathrm{Re} = \frac{cd_1}{\nu}, \quad Da = \frac{\bar{K}}{d_1^2}, \quad \theta = \frac{T - T_0}{T_1 - T_0}, \quad \mathrm{Pr} = \frac{\rho v \xi}{k}, \quad E = \frac{c^2}{\xi(T_1 - T_0)}$$

,

$$M^2 = \frac{\sigma B_0^2 d_1^2}{\mu}, \quad F_r = \frac{c^2}{g d_1}, \quad u = \frac{\partial \psi}{\partial y}, \quad v = -\frac{\partial \psi}{\partial x}.$$

Using the above non-dimension quantities the continuity equation is satisfied and equations (5.5) to (5.7) under long wave length and low Reynolds number assumption reduce to

$$\frac{dp}{dx} = \frac{\partial^3 \psi}{\partial y^3} - \left(\frac{1}{Da} + M^2 \right) \left(\frac{\partial \psi}{\partial y} + 1 \right) + \frac{R}{F_r} \sin \alpha, \qquad (5.9)$$

$$\frac{1}{\mathrm{Pr}} \left(\frac{\partial^2 \theta}{\partial y^2} \right) + E \left(\frac{\partial^2 \psi}{\partial y^2} \right)^2 = 0, \qquad (5.10)$$

Equation (5.9) can be rewritten as

$$\frac{\partial^4 \psi}{\partial \psi^4} - \left(\frac{1}{Da} + M^2 \right) \frac{\partial^2 \psi}{\partial \psi^2} = 0 \quad . \qquad (5.11)$$

The boundary conditions are

$$\psi = \frac{q}{2}, \quad y = h_1(x) = 1 + a\cos 2\pi x, \qquad (5.12)$$

$$\psi = \frac{-q}{2}, \quad y = h_2(x) = -d - b\cos(2\pi x + \phi), \qquad (5.13)$$

$$\frac{\partial \psi}{\partial y} = -1 \quad at \ y = h_1 \ and \ y = h_2, \qquad (5.14)$$

201

$\theta = 0$ *at* y=h_1 and θ=1 at y=h_2. (5.15)

and $a^2 + b^2 + 2ab\cos\phi \leq (1+d)^2$. Here $\partial p / \partial y = 0$, ψ is the stream function, q is the flux in the wave frame, M is the Hartman number, Pr and E are Prandtl and Eckert numbers, respectively.

5.4. Solution of the problem

The solutions for equation (5.11) satisfying the corresponding boundary conditions (5.12) to (5.14) are

$$\Psi = C_1 y + C_2 + C_3 \cosh Ny + C_4 \sinh Ny.$$ (5.16)

The velocity is given by

$$u = C_1 + NC_3 \sinh Ny + NC_4 \cosh Ny.$$ (5.17)

where

$$N^2 = \frac{1}{Da} + M^2,$$

$$C_1 = \frac{Nq + 2\tanh\left[\dfrac{N(h_1 - h_2)}{2}\right]}{N(h_1 - h_2) - 2\tanh\left[\dfrac{N(h_1 - h_2)}{2}\right]},$$

$$C_2 = \frac{(h_1 + h_2)\left[Nq + 2\tanh\left[\dfrac{N(h_1 - h_2)}{2}\right]\right]}{2N(h_2 - h_1) + 4\tanh\left[\dfrac{N(h_1 - h_2)}{2}\right]},$$

$$C_3 = \frac{(q + h_1 - h_2)\sec h\left[\dfrac{N(h_1 - h_2)}{2}\right]\sinh\left[\dfrac{N(h_1 + h_2)}{2}\right]}{N(h_1 - h_2) - 2\tanh\left[\dfrac{N(h_1 - h_2)}{2}\right]},$$

$$C_4 = \frac{(q + h_1 - h_2)\sec h\left[\dfrac{N(h_1 - h_2)}{2}\right]\cosh\left[\dfrac{N(h_1 + h_2)}{2}\right]}{N(h_2 - h_1) + 2\tanh\left[\dfrac{N(h_1 - h_2)}{2}\right]}.$$

Making use of equation (5.10), the solution of equation of equation (5.11) satisfying the boundary conditions (5.15) is given by

$$
\theta = -\frac{Br(N^2)}{2}
\begin{bmatrix}
\left(\dfrac{C_3^2 - C_4^2}{2}\right)\left\{(y^2 - h_1^2) - (h_2 + h_1)(y - h_1)\right\} \\[2mm]
+\dfrac{(C_3^2 + C_4^2)}{4N}\left\{(\cosh[2Ny] - \cosh[2Nh_1]) - \dfrac{(y - h_1)}{(h_2 - h_1)}C_5\right\} \\[2mm]
+\dfrac{C_3 C_4}{2N}\left\{(\sinh[2Ny] - \sinh[2Nh_1]) - \dfrac{(y - h_1)}{(h_2 - h_1)}C_6\right\}
\end{bmatrix}
+\frac{(y - h_1)}{(h_2 - h_1)}. \quad (5.18)
$$

where $Br = E\,Pr$ is the

$$C_5 = \cosh[2Nh_2] - \cosh[2Nh_1],$$

$$C_6 = \sinh[2Nh_2] - \sinh[2Nh_1].$$

The dimensionless flux in the fixed frame is given by

$$
Q = \int_{h_2}^{h_1}(u + 1)dy = q + h_1 - h_2. \quad (5.19)
$$

The average volume flow rate over one period ($T = \dfrac{\lambda}{c}$) of the peristaltic wave is given by

$$
\bar{Q} = \frac{1}{T}\int_0^T Qdt = q + 1 + d. \quad (5.20)
$$

The pressure gradient is obtained as

$$
\frac{dp}{dx} = -N^3(C_1 + 1) + \frac{R_e}{F_r}\sin\alpha. \quad (5.21)
$$

The pressure rise per wave length is denoted by ΔP_λ and is given by

$$
\Delta P_\lambda = \int_0^1 \frac{dp}{dx}\,dx. \quad (5.22)
$$

The frictional forces, at $y = h_1$ and $y = h_2$ denoted by $F_{\lambda 1}$ and $F_{\lambda 2}$, respectively are given as follows

$$F_{\lambda 1} = \int_0^1 -h_1^2 \left(\frac{dp}{dx} \right) dx \,, \tag{5.23}$$

$$F_{\lambda 2} = \int_0^1 -h_2^2 \left(\frac{dp}{dx} \right) dx \,. \tag{5.24}$$

5.5 Expressions for wave shape:

We can deduce the symmetric channel by taking a=b, d=1 and ϕ=0. The non-dimensional expressions for the three considered wave forms are given by the following equations:

1. Sinusoidal wave:

$$h(x) = 1 + a\sin(x), \tag{5.25}$$

2. Triangular wave:

$$h(x) = 1 + a \left\{ \frac{8}{\pi^3} \sum_{m=1}^{\infty} \frac{(-1)^{m+1}}{(2m-1)^2} \sin[(2m-1)x] \right\}, \tag{5.26}$$

3. Trapezoidal wave:

$$h(x) = 1 + a \left\{ \frac{32}{\pi^2} \sum_{m=1}^{\infty} \frac{\sin \frac{\pi}{8}(2m-1)}{(2m-1)^2} \sin[(2m-1)x] \right\}. \tag{5.27}$$

5.6 Results and discussion:

When the angle of inclination α is taken as zero, the results deduced agree with the corresponding ones of Hayat (2009).

The variation of velocity u with y is calculated from equation (5.17) for different values of Hartmann number M with x=1, a=0.5, b=0.6, d=1, $\phi = \frac{\pi}{6}$, $\overline{Q}$ =1and Da =0.1 and is depicted in figure (5.2). It is observed that velocity profiles are parabolic for fixed values of the Hartmann number M. It is also observed that the velocity 'u' increases with increasing Hartmann number M near the walls. However, u decreases by increasing M near the centre of the channel.

The variation between velocity u with y is drawn in figure (5.3) for different values of the phase difference ϕ with x=1, a=0.5, b=0.6, d=1, M=1, $\overline{Q}$=1, and Da

=0.1. It is concluded that the velocity decreases with increasing phase difference ϕ in the lower half of the channel and increasing on the upper half of the channel.

Figure (5.4) shows that the relation between velocity u with y for various values of Darcy number Da with a=0.5, b=0.5, d=1, M=1, $\phi = \frac{\pi}{3}$ and $\overline{Q}$=1. It also observed that the velocity 'u' decreases with increasing Darcy number Da near the walls. However, u increases by increasing Da near the centre of the channel.

The relation between velocity u with y is depicted in figure (5.5) for different the average volume flow rate with x=1, a=0.5, b=0.6, d=1, M=1, $\phi = \frac{\pi}{3}$ and Da =0.1. It is concluded that the velocity u increases with increasing the average volume flow rate $\overline{Q}$.

The variation of pressure gradient $\frac{dp}{dx}$ with x is calculated from equation (5.21) for different values of Darcy number Da with a=0.5, b=0.5, d=1, $\phi = 0$, $\overline{Q}$ =1, $\phi = \frac{\pi}{6}$, R_e=0.1, F_r=0.1 and $\alpha = \frac{\pi}{4}$ and is plotted in figure (5.6). It is observed that pressure gradient $\frac{dp}{dx}$ decreases with increasing Darcy number Da .

In figure (5.7) the variation of pressure gradient $\frac{dp}{dx}$ with x is drawn for different values of phase difference ϕ with x=1, a=0.5, b=0.5, $\phi = \frac{\pi}{6}$, d=1, M=1, $\overline{Q}$ =1, R_e=0.1, F_r=0.2, Da =0.1 and $\alpha = \frac{\pi}{4}$. It is observed that the pressure gradient $\frac{dp}{dx}$ decreases with increasing phase angle ϕ.

The relation between pressure gradient $\frac{dp}{dx}$ with x is plotted in figure (5.8) for different values of Hartmann number M with a=0.5, b=0.5, d=1, $\overline{Q}$=1, R_e=0.2, F_r=0.1, Da =0.1 and $\alpha = \frac{\pi}{4}$. It is found that the pressure gradient $\frac{dp}{dx}$ increases with increasing Hartmann number M.

In figure (5.9) the relation between pressure gradient $\dfrac{dp}{dx}$ and x is plotted at different values of the average volume flow rate $\overline{Q}$ with a=0.5, b=0.5, d=1, R_e=0.2, F_r=0.1, M=1,$\phi = \dfrac{\pi}{6}$, Da=0.1 and $\alpha = \dfrac{\pi}{4}$. It is noticed that the pressure gradient $\dfrac{dp}{dx}$ increases with increasing the average volume flow rate $\overline{Q}$.

The variation of the pressure gradient $\dfrac{dp}{dx}$ with x is depicted in figure (5.10) for various values of Reynolds number R_e with a=0.5, b=0.5, d=1, R_e=0.2, F_r=0.1, M=1,$\phi = \dfrac{\pi}{6}$, Da=0.1 and $\alpha = \dfrac{\pi}{4}$. It is noticed that the pressure gradient$\dfrac{dp}{dx}$ increases with increasing Reynolds number R_e.

The relation between pressure gradient $\dfrac{dp}{dx}$ with x is shown in figure (5.11) for different values of Froude number F_r with a=0.5, b=0.5, d=1,$\phi = \dfrac{\pi}{6}$, $\overline{Q}$=1, R_e =0.1, Da=0.1, M=1, and $\alpha = \dfrac{\pi}{4}$. It is concluded that the pressure gradient $\dfrac{dp}{dx}$ decreases with increasing Froude number F_r.

The relation between pressure gradient $\dfrac{dp}{dx}$ and x is drawn in figure (5.12) for different values of inclined angle α with a=0.5, b=0.5, d=1, $\overline{Q}$=1, M=1,$\phi = \dfrac{\pi}{6}$, R_e=0.1, F_r=0.1 and Da=0.1. It is found that the pressure gradient $\dfrac{dp}{dx}$ increases with increasing angle of inclination α $\left(0 \le \alpha \le \dfrac{\pi}{2}\right)$.

From figure (5.13) the variation pressure gradient $\dfrac{dp}{dx}$ with x is plotted for different channel width d with a=0.5, b=0.5, $\overline{Q}$=1, M=1,$\phi = \dfrac{\pi}{6}$, R_e=0.1, F_r=0.1, $\alpha = \dfrac{\pi}{4}$ and Da=0.1. It is concluded that the pressure rise deceases by increasing channel width d.

The variation on pressure rise ΔP_λ with the average flow rate $\overline{Q}$ is calculated from equation (5.22) and is plotted figure (5.14) for different values of Darcy number Da with a=0.5, b=0.5, d=1, M=1, $\phi = \frac{\pi}{6}$, R_e=0.1, $\alpha = \frac{\pi}{4}$ and F_r=0.1. We conclude that for values of $\overline{Q}$ between 0.2 and 0.3, the pumping curves intersect at a point (0.24, 0.6). For a given mean flux $\overline{Q}$, the pressure rise ΔP_λ increases with increasing Darcy number Da below this point and opposite behaviour is observed above this point.

Figure (5.15) shows the relation between pressure rise ΔP_λ with the average flow rate $\overline{Q}$ for various values of Froude number F_r with a=0.5, b=0.5, d=1, M=1, $\phi = \frac{\pi}{6}$, R_e=0.1, $\alpha = \frac{\pi}{4}$ and Da=1. It is observed pressure rise ΔP_λ decreases with increasing the Froude number F_r.

In figure (5.16) shows the relation between pressure rise ΔP_λ with the average flow rate $\overline{Q}$ for various values of inclined angle α with a=0.5, b=0.5, d=1, M=1, $\phi = \frac{\pi}{6}$, R_e=0.1, F_r=0.1 and Da=1.. It is observed pressure rise ΔP_λ increases with increasing an angle of inclination α $\left(0 \leq \alpha \leq \frac{\pi}{2}\right)$. We also infer obtained in the case of a vertical channel than in the case of a horizontal channel.

The relation between pressure rise ΔP_λ with the average flow rate $\overline{Q}$ is plotted figure (5.17) for different values of Hartmann number M with a=0.5, b=0.5, d=1, $\phi = \frac{\pi}{6}$, R_e=0.1, F_r=0.1, $\alpha = \frac{\pi}{4}$ and Da=1. We observe that for values of $\overline{Q}$ between 0.2 and 0.3 the pumping curves intersect at a point (0.24, 0.65). For a given the average flow rate $\overline{Q}$, the pressure rise ΔP_λ decreases with increasing Hartmann number M below this point and opposite behaviour is observed above this point.

The relationship between pressure rise ΔP_λ with the average flow rate $\bar{Q}$ is drawn figure (5.18) for different values of Reynolds number R_e with a=0.5, b=0.5, d=1, M=1, $\phi = \dfrac{\pi}{6}$, F_r=0.1, $\alpha = \dfrac{\pi}{4}$ and Da=1. We conclude that the pressure rise ΔP_λ increases with increasing Reynolds number R_e.

The variation of frictional forces $F_{\lambda 1}$ and $F_{\lambda 2}$ with $\bar{Q}$ is shown in figures (5.19) to (5.28). We observe that the frictional force shows opposite behaviour to that of pressure rise for the corresponding variations in the physical parameters M, ϕ , R_e, F_r, α and Da .

The variation on temperature θ with y is calculated from equation (5.18) and is drawn in figure (5.29) for different values of average flow rate $\bar{Q}$ with x=0, a=0.7, b=1.2, d=2, M=1, ϕ=0, Da=1and Br=1. We conclude that the temperature distribution θ increases with increasing $\bar{Q}$ at inlet of the channel.

The relation between the temperature θ with y is plotted in figure (5.29) for different values of Brinkman number Br with x=0, a=0.7, b=1.2, d=2, M=1, ϕ=0, Da=1 and $\bar{Q}$=1. We observed that the temperature θ increases with increasing Brinkman number Br.

Figure (5.31) the variation of the temperature θ with y for different values of Hartmann number M with x=0, a=0.7, b=1.2, d=2, ϕ=π/3, Br=1, Da=1and $\bar{Q}$=1. It is noticed that increasing with M the temperature θ decreases near the lower wall and increases with M in the centre of the channel.

The relationship between the temperature θ with y in figure (5.32) for different values of Darcy number Da with x=0, a=0.7, b=1.2, d=2, ϕ=π/3, Br=1, M=1, and $\bar{Q}$=1. It is concluded that the temperature θ decreases with increasing Darcy number Da .

5.6.1 Trapping phenomena

Another interesting phenomenon in peristaltic motion is trapping. It is basically the formation of an internally circulating bolus of fluid by closed stream lines. The trapped bolus will be pushed ahead along the peristaltic waves.

The stream lines are plotted in figure (5.33) for various values of Hartmann number M with a=0.5, b=0.5, d=1, ϕ=0, $\bar{Q}$=1 and Da =0.2. It is concluded that the volume of the trapping bolus decreases with increasing Hartmann number M and the bolus disappears for M=8.

The stream lines are depicted in figure (5.34) for different values of volume flow rate $\bar{Q}$ with a=0.5, b=0.5, d=1, M=1, ϕ=0, and Da =0.1. It is noticed that the size of the trapping bolus increases with increasing the average volume flow rate $\bar{Q}$ and the bolus disappears for $\bar{Q}$=1.2.

The stream lines are drawn in figure (5.35) for different values of Darcy number Da with a=0.5, b=0.5, d=1, M=1, ϕ=0 and $\bar{Q}$=1. It is found that the volume of the trapping bolus increases with increasing Darcy number Da and the bolus disappears for Da =0.01.

The effects of phase difference ϕ on trapping with same amplitudes a=0.5, b=0.5, d=1, M=1, Da =0.1 and $\bar{Q}$=1 is plotted in figure (5.36). It is observed that the bolus appearing in the central region for ϕ=0 moves towards left and decreases in size as ϕ increases. For ϕ=π the bolus disappears and stream lines are parallel to the boundary walls.

Stream lines are plotted in figures (5.37) and (5.38) for different values of the average volume flow rate $\bar{Q}$ with a=0.5, b=0.5, d=1, ϕ=0, Da =0.1 and M=1for following wave forms a) sinusoidal wave b) square wave and c) triangular wave. We observed that the size of the bolus increases with increasing $\bar{Q}$ for the wave forms considered.

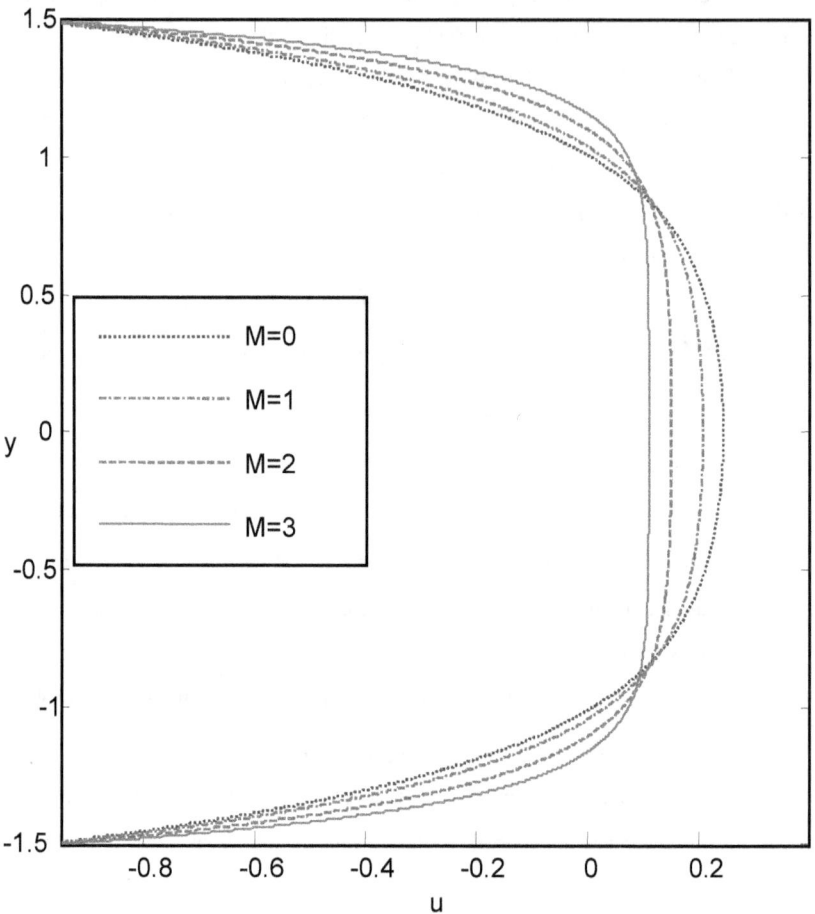

Figure 5.2: The velocity profiles with a=0.5, b=0.6, d=1, x=1, $\phi = \dfrac{\pi}{6}$, $\overline{Q}$=1 and Da =0.1.

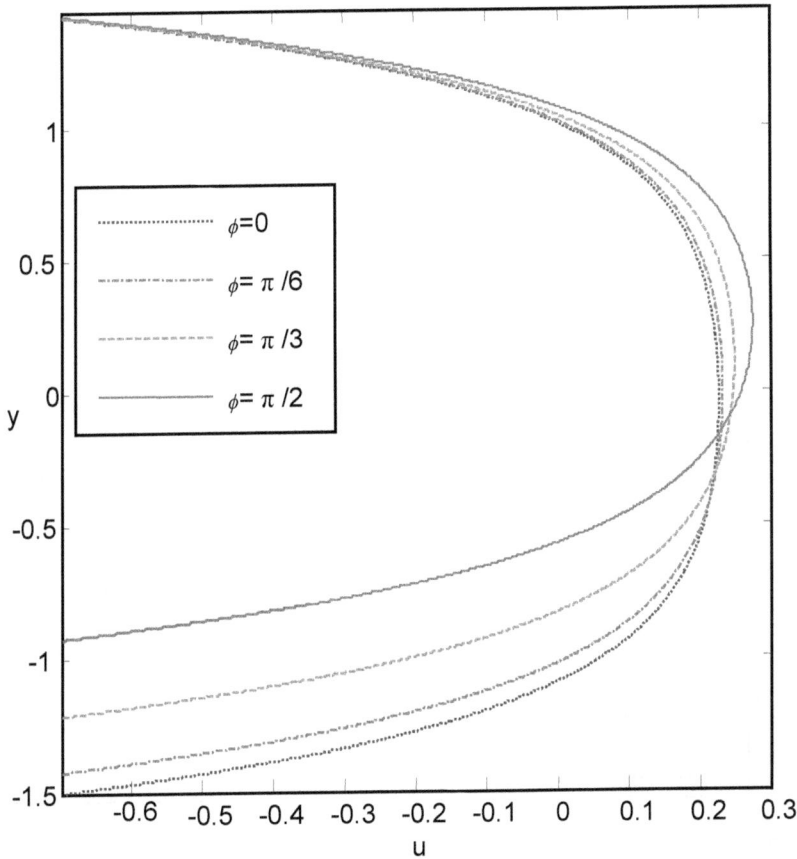

Figure 5.3: The velocity profiles with a=0.5, b=0.6, d=1, x=1, M=1, $\bar{Q}$=1 and Da
=0.1.

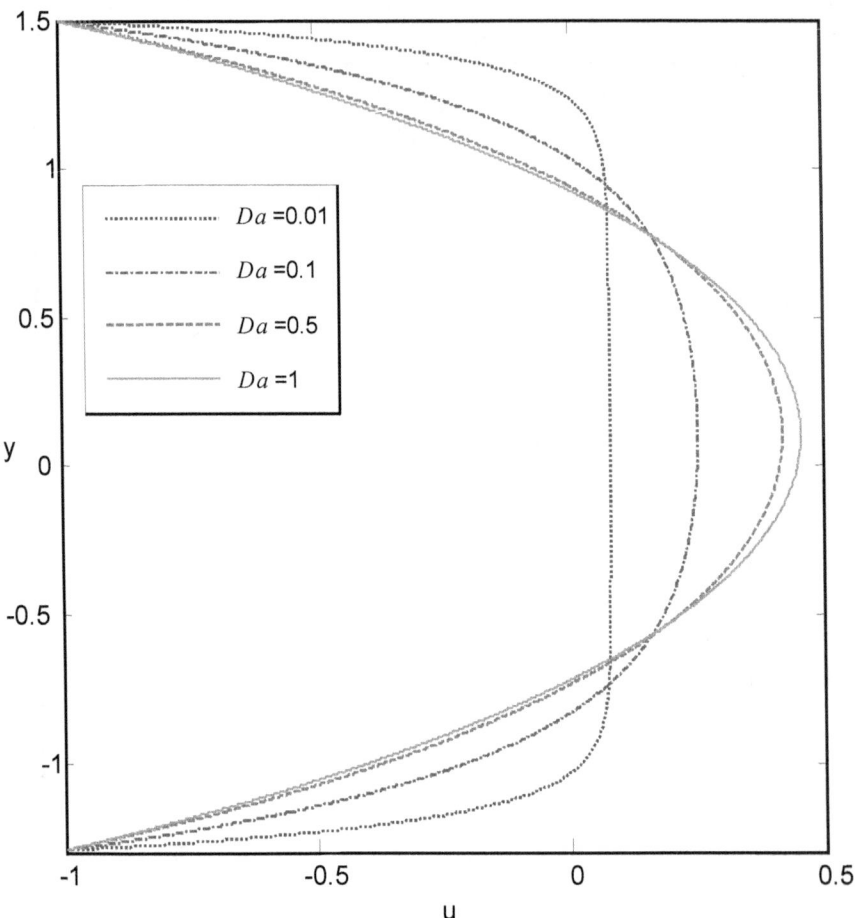

Figure 5.4: The velocity profiles with a=0.5, b=0.6, d=1, x=1, M=1, $\phi = \dfrac{\pi}{3}$ and $\overline{Q}$

=1.

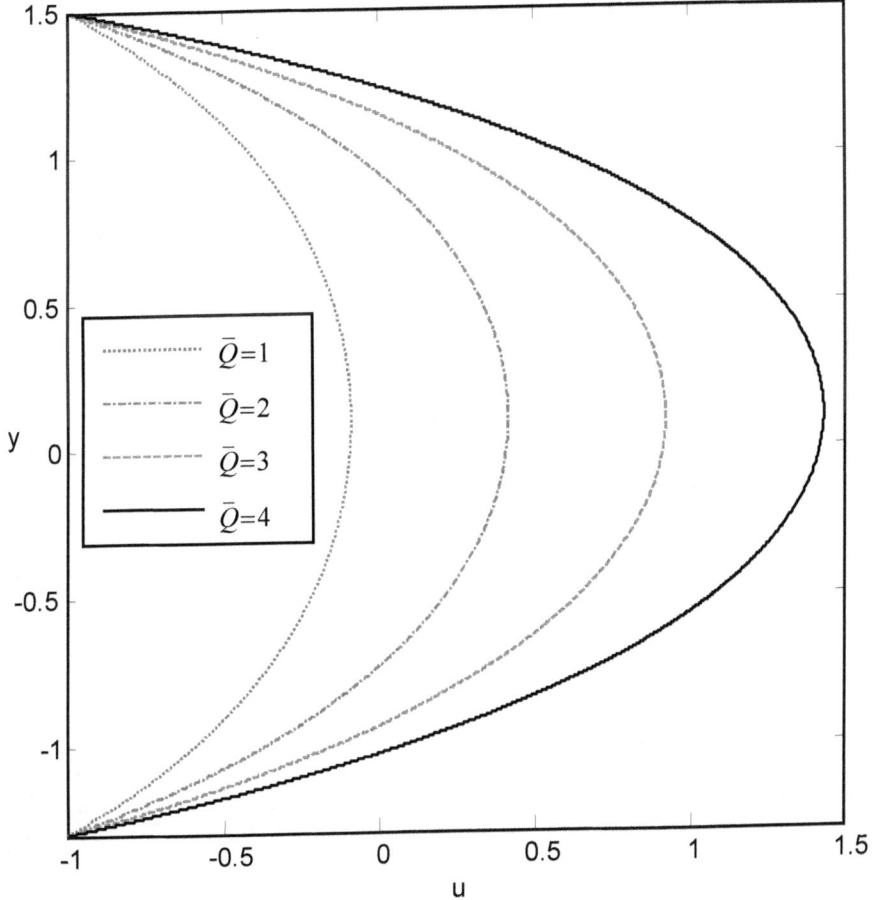

Figure 5.5: The velocity profiles with a=0.5, b=0.6, d=1, x=1, M=1, $\phi = \frac{\pi}{3}$ and Da =0.1.

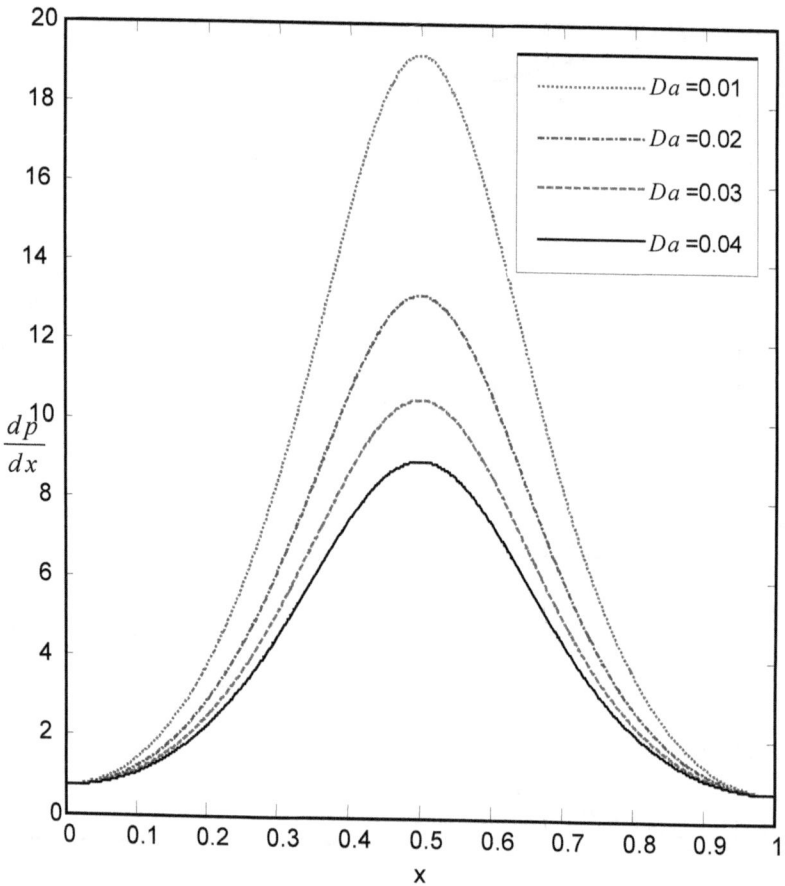

Figure 5.6: Various of pressure gradient $\dfrac{dp}{dx}$ with x for different values of Darcy

number Da with a=0.5, b=0.5, d=1, M=1, $\phi = 0$ $\overline{Q}$=1, R_e=0.1, F_r=0.1 and

$\alpha = \dfrac{\pi}{4}$.

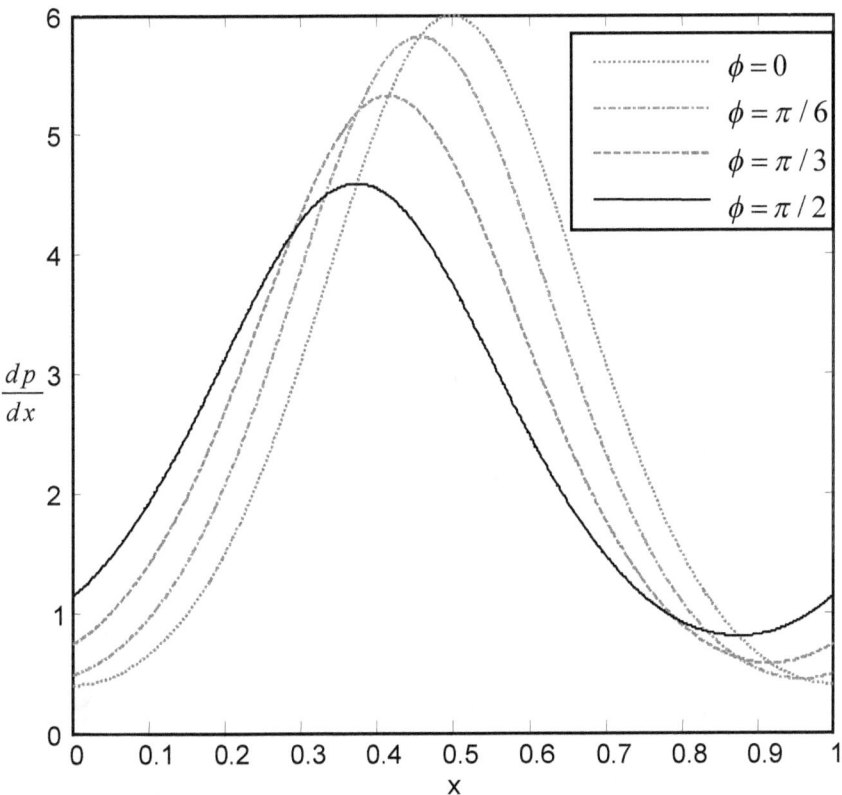

Figure 5.7: Various of pressure gradient $\dfrac{dp}{dx}$ with x for different value of phase

difference ϕ with   a=0.5, b=0.5, d=1, M=1, $\overline{Q}$=1, Da=0.1, R_e=0.2,

$\alpha = \dfrac{\pi}{4}$ and F_r=0.1.

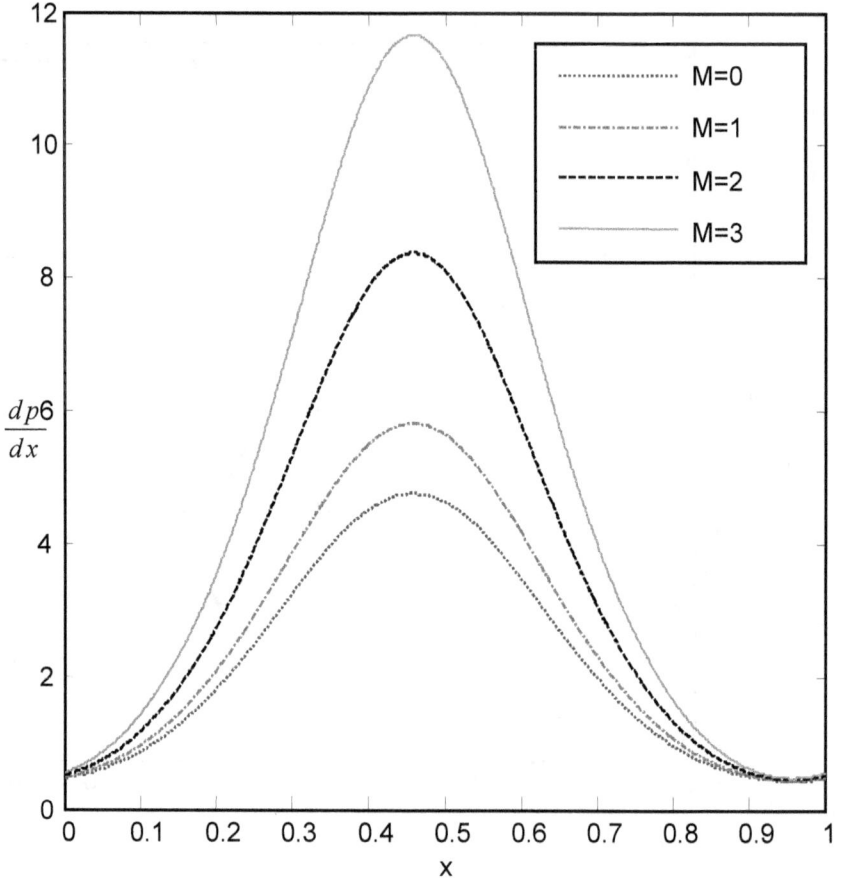

Figure 5.8: Various of pressure gradient $\dfrac{dp}{dx}$ with x for different values of Hartmann

number M with   a=0.5, b=0.5, d=1, $\overline{Q}=1, \phi = \pi / 6, Da =0.1,$ $R_g=0.2,$

$\alpha = \dfrac{\pi}{4}$ and $F_r =0.1$.

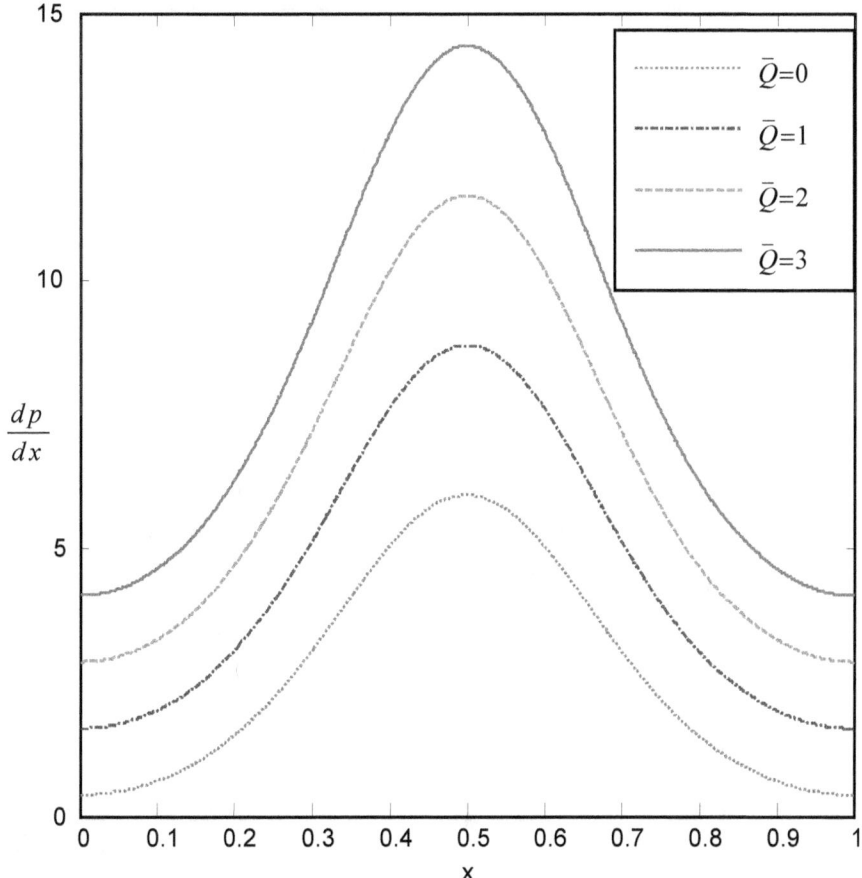

Figure 5.9: Various of pressure gradient $\dfrac{dp}{dx}$ with x for different values average

volume floe rate of $\overline{Q}$ with   a=0.5, b=0.5, d=1, M=1, $\phi = \pi/6$, Da =0.1,

R_e=0.1, $\alpha = \dfrac{\pi}{4}$ and F_r =0.1.

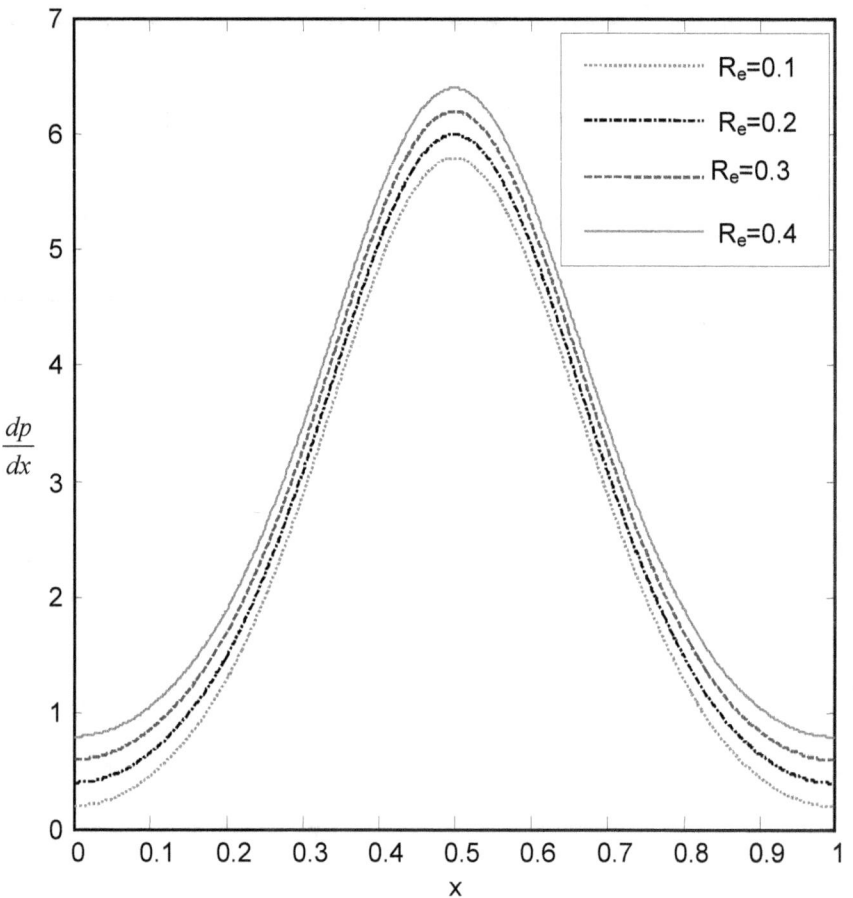

Figure 5.10: Various of pressure gradient $\frac{dp}{dx}$ with x for different values of

Reynolds number R_e with a=0.5, b=0.5, d=1, $\overline{Q}$=1, M=1, $\phi = \pi/6$, Da

=0.1, $\alpha = \frac{\pi}{4}$ and $F_r = 0.1$.

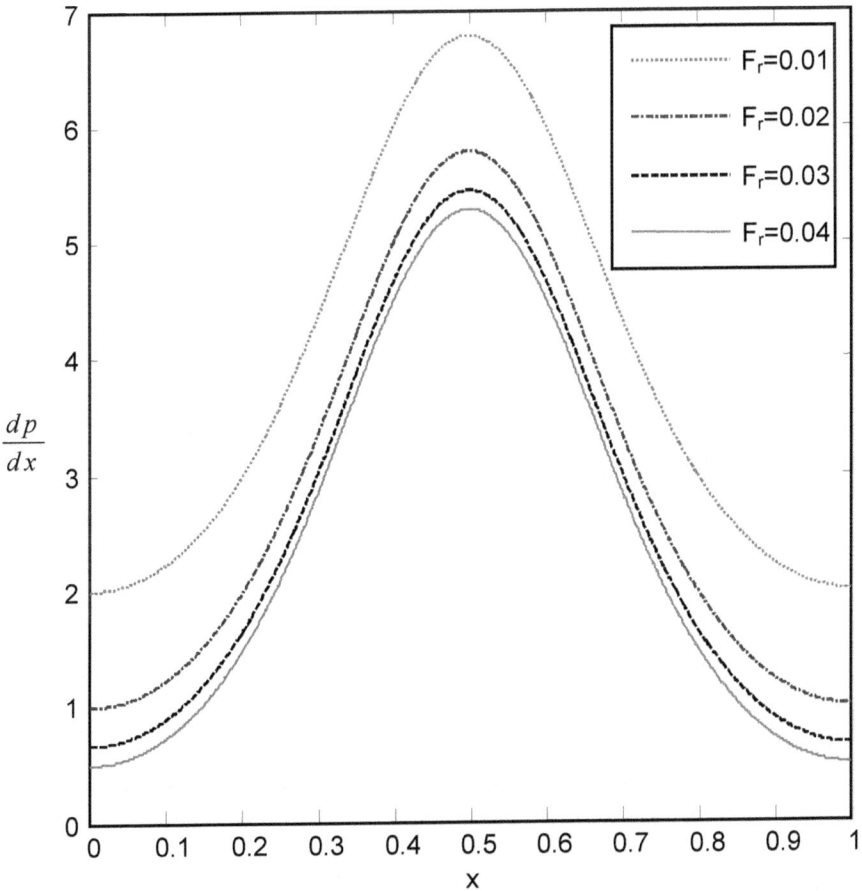

Figure 5.11: Various of pressure gradient $\dfrac{dp}{dx}$ with x for different values of Froud

number F_r with   a=0.5, b=0.5, d=1, $\overline{Q}$=1, M=1, $\phi=\pi/6$, Da =0.1, R_v

=0.1 and $\alpha = \dfrac{\pi}{4}$.

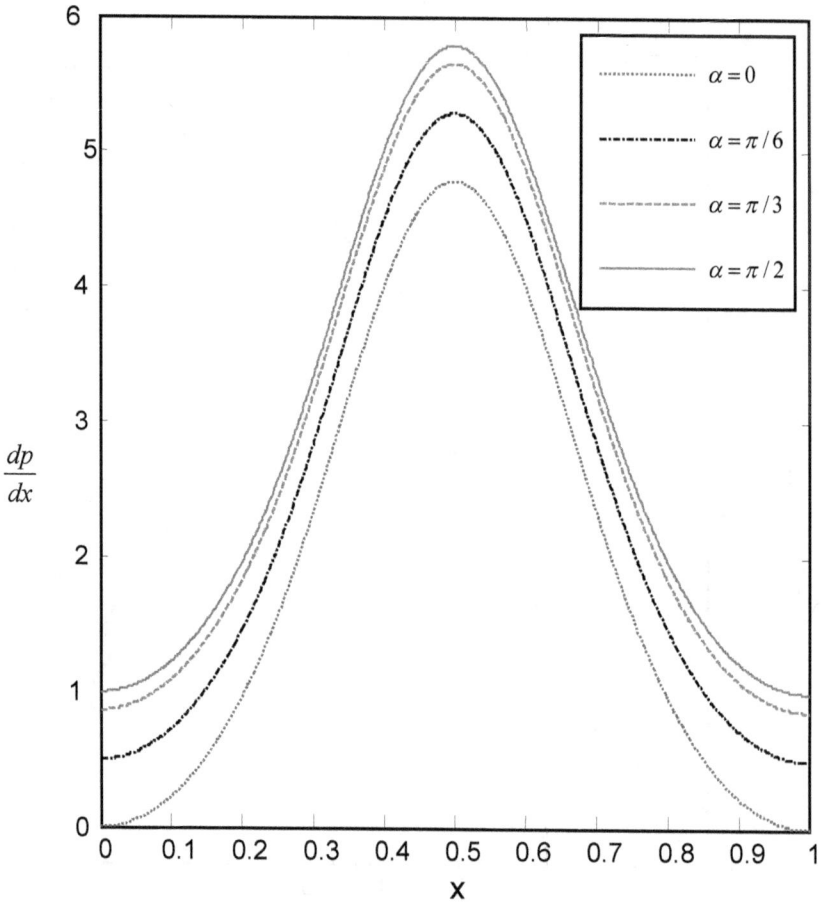

Figure 5.12: Various of pressure gradient $\dfrac{dp}{dx}$ with x for different values of angle of

inclination α with a=0.5, b=0.5, d=1, $\overline{Q}$=1, M=1, $\phi = \pi/6$, Da =0.1, R_e

=0.1 and F_r =0.1.

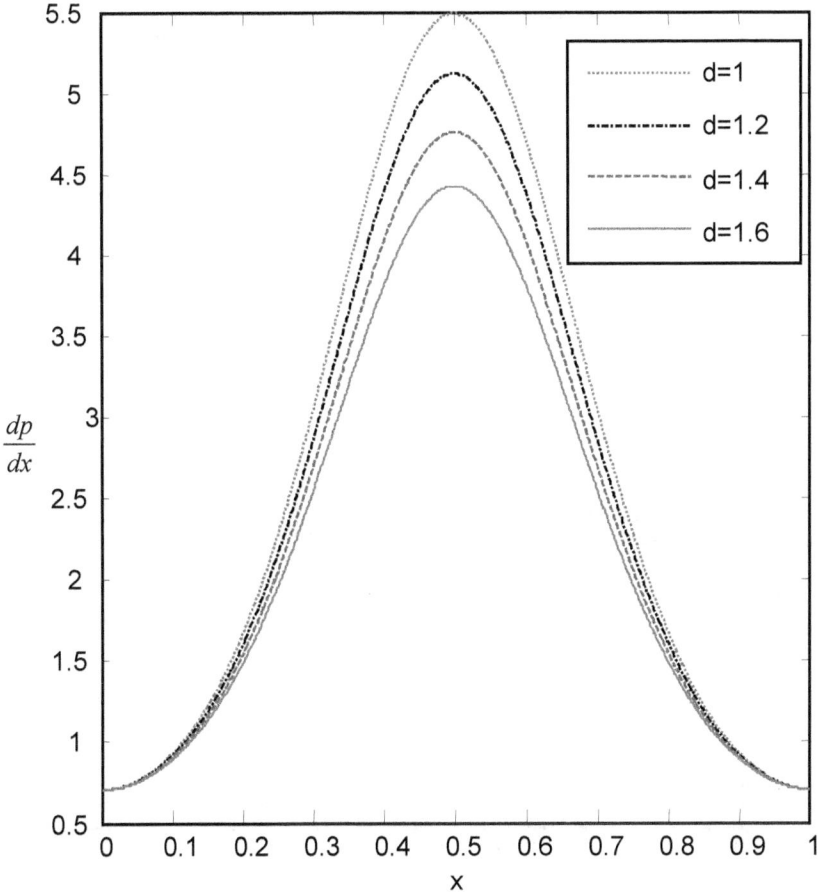

Figure 5.13: Various of pressure gradient $\frac{dp}{dx}$ with x for different values of

channel width d with   a=0.5, b=0.5, $\overline{Q}$=1, M=1, $\phi = \pi / 6$, Da =0.1,

R_e=0.1, $\alpha = \frac{\pi}{4}$ ϕ = π/6 and F_r =0.1.

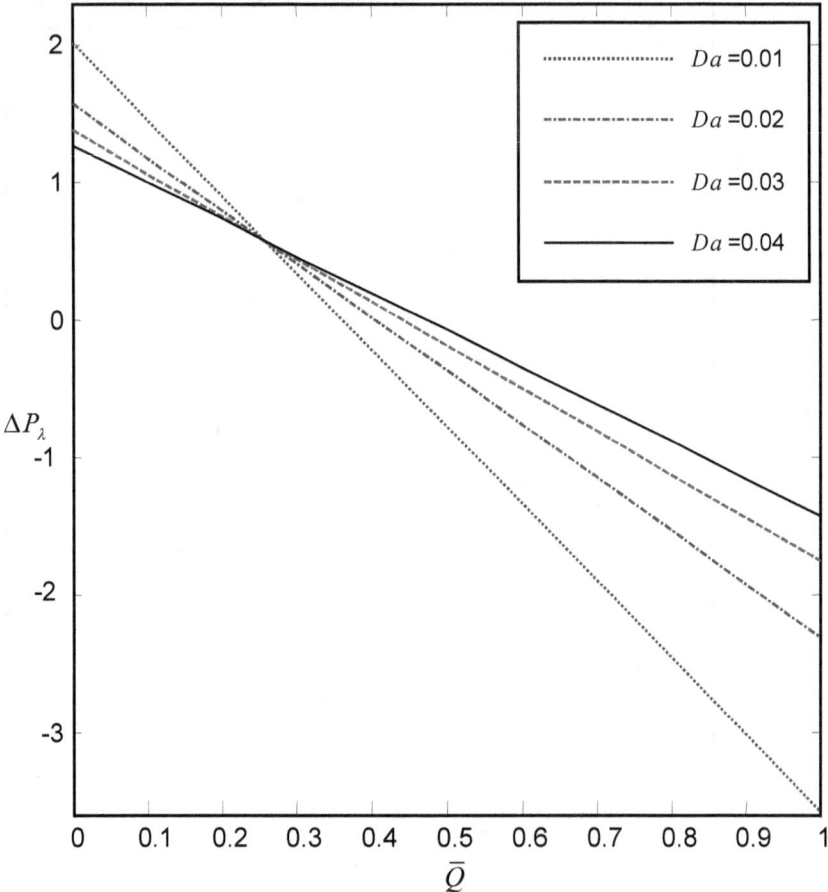

Figure 5.14: Various of $\bar{Q}$ with ΔP_λ for different values of Darcy number Da with

a=0.5, b=0.5, d=1, M=1, $\phi = \pi / 6$, R_e =0.1, $\alpha = \dfrac{\pi}{4}$ and $F_r = 0.1$

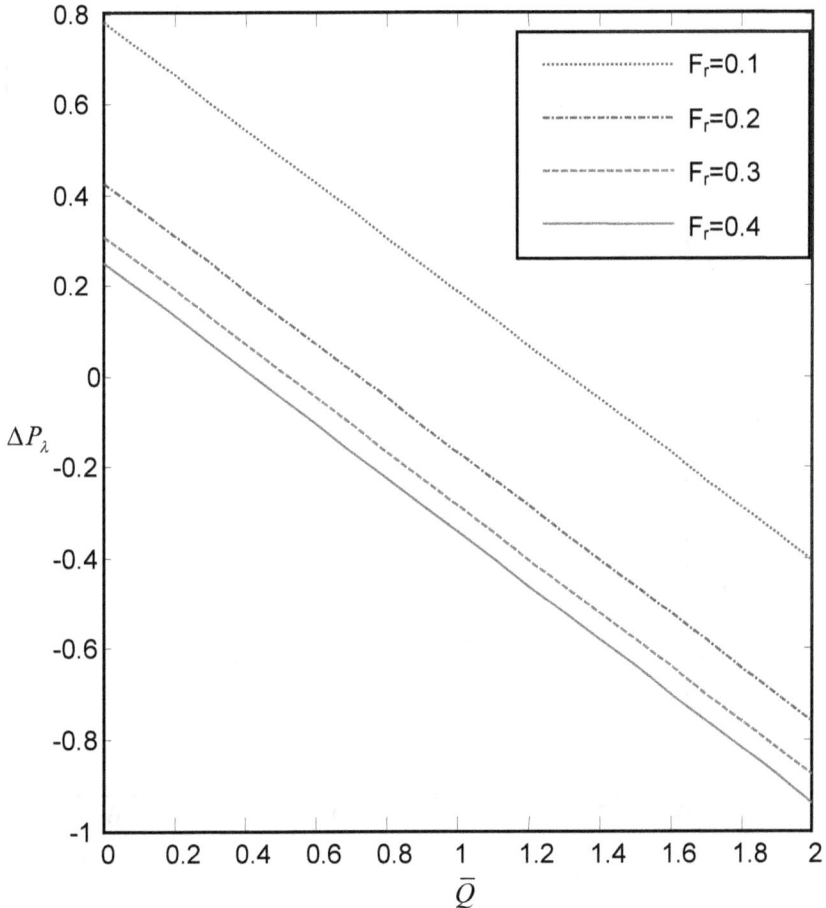

Figure 5.15: Various of $\overline{Q}$ with ΔP_λ for different values of Froud number Fr with

a=0.5, b=0.5, d=1, M=1, $\phi = \pi / 6$, Da =0.1, R_e =0.1 and $\alpha = \dfrac{\pi}{4}$

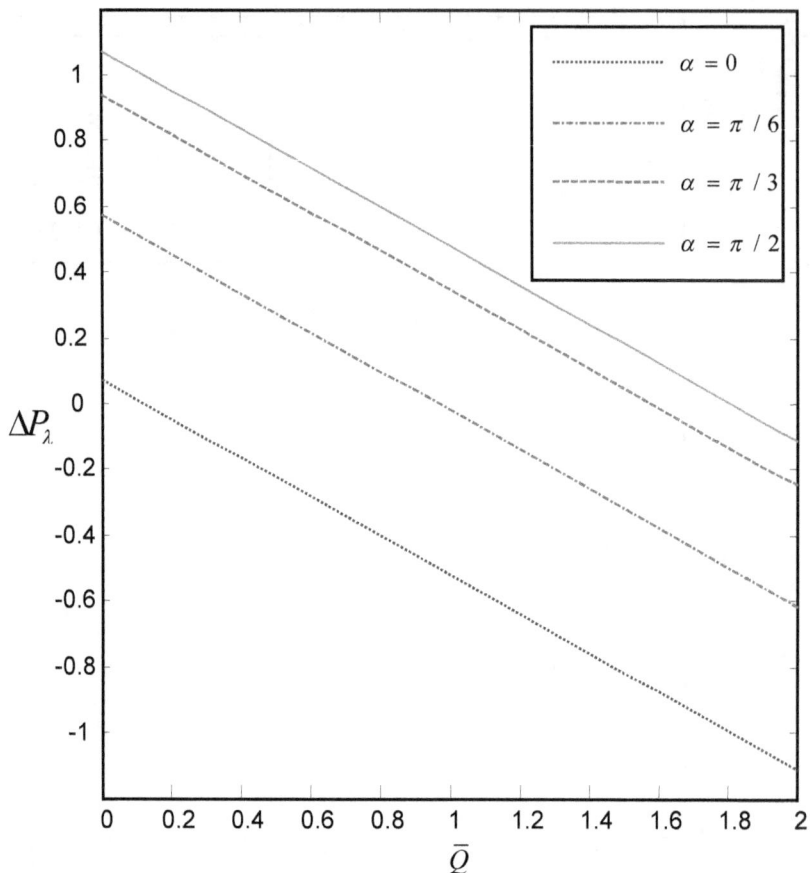

Figure 5.16: Various of $\overline{Q}$ with ΔP_λ for different values of angle of inclination α

with a=0.5, b=0.5, d=1, M=1, $\phi = \pi / 6$, Da =0.1, R_e =0.1, and F_r =0.1

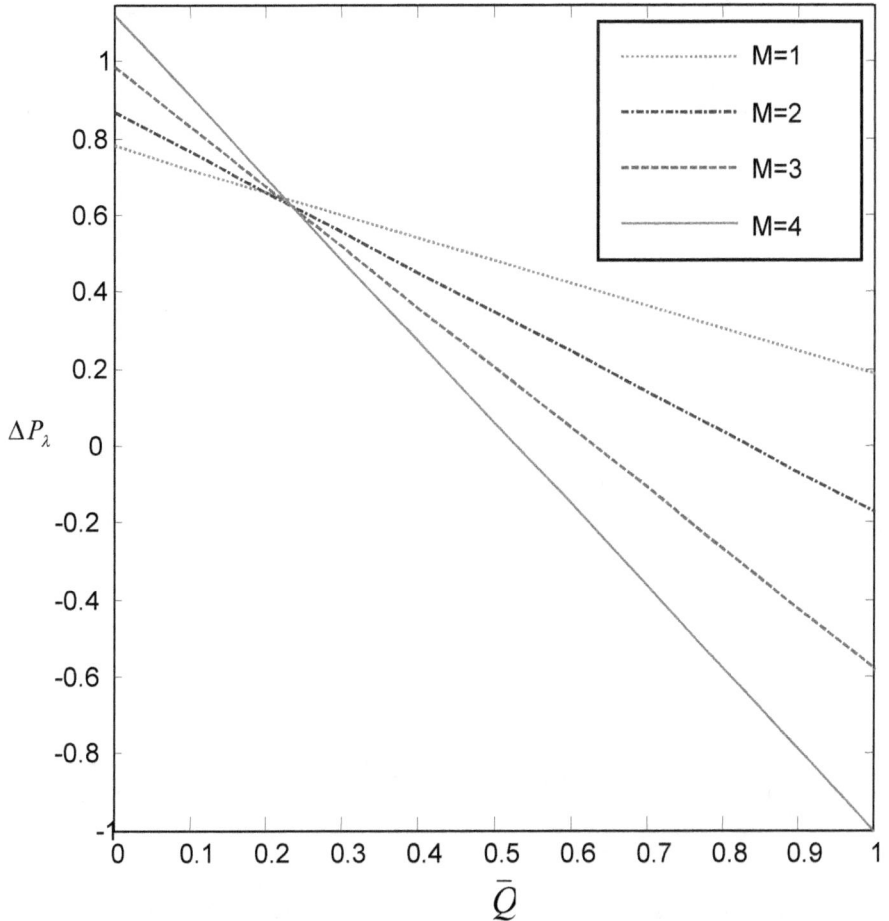

Figure 5.17: Various of $\overline{Q}$ with ΔP_λ for different values of Hartmann number M

with a=0.5, b=0.5, d=1, $\phi = \pi / 6$, Da =0.1, R_e =0.1, α =π/4 and F_r =0.1

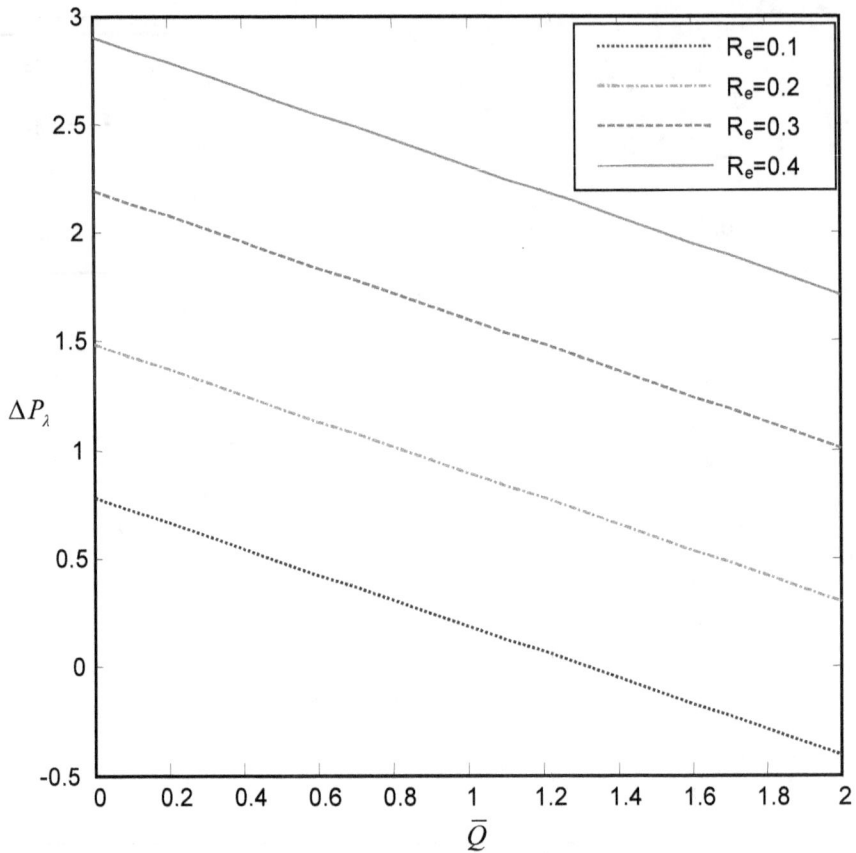

Figure 5.18: Various of $\bar{Q}$ with ΔP_λ for different values of Reynolds number R_e with a=0.5, b=0.5, d=1, M=1, $\phi = \pi/6$, $Da = 0.1$, $\alpha = \pi/4$ and $F_r = 0.1$.

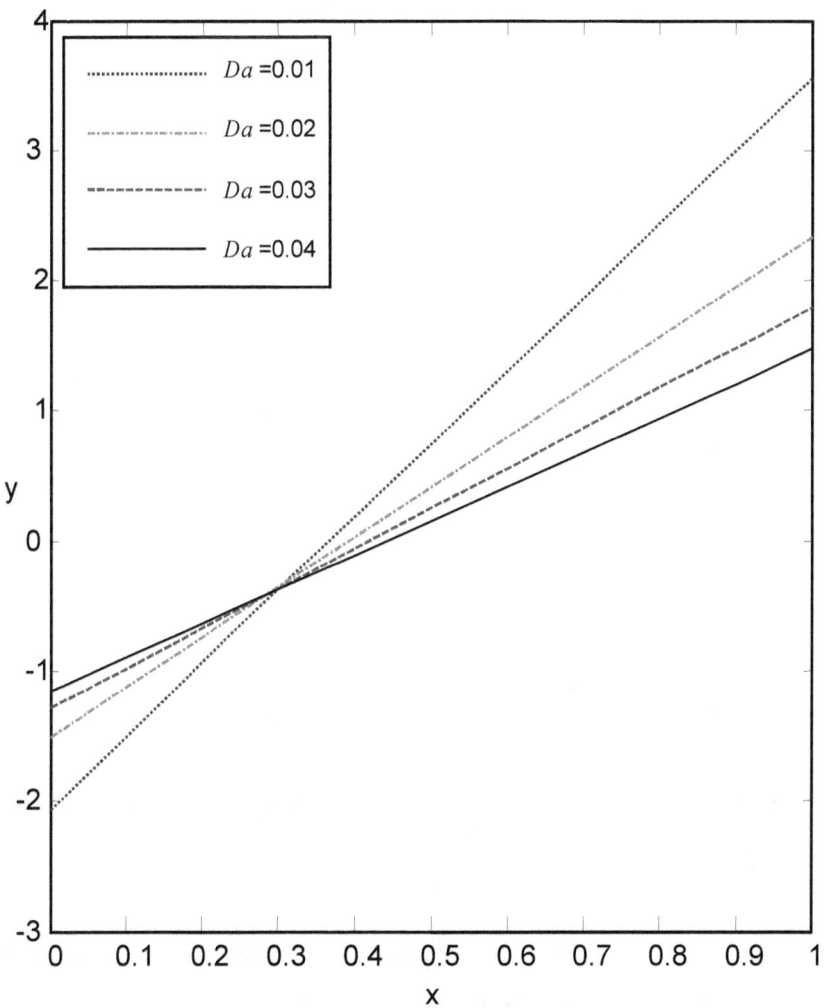

Figure 5.19: Various of $\overline{Q}$ with $F_{\lambda 1}$ at $y=h_1$ for different values of Darcy number Da with a=0.2, b=1.2, d=1, M=1, $\phi=\pi/6$, $\alpha=\pi/4$, $R_e=0.1$ and $F_r=0.1$.

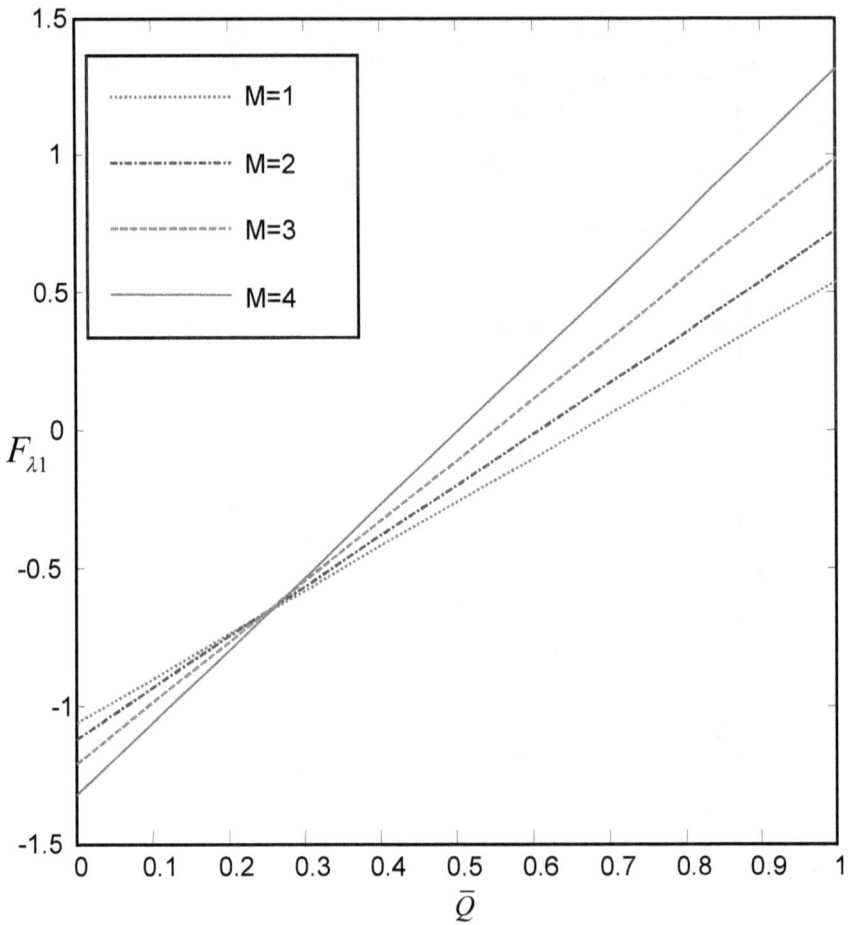

Figure 5.20: Various of $\overline{Q}$ with $F_{\lambda 1}$ at $y=h_1$ for different values of Hartmann number M with a=0.2, b=1.2, d=1, $\phi=\pi/6$, Da=0.1, α=π/6, R_e=0.1 and F_r=0.1.

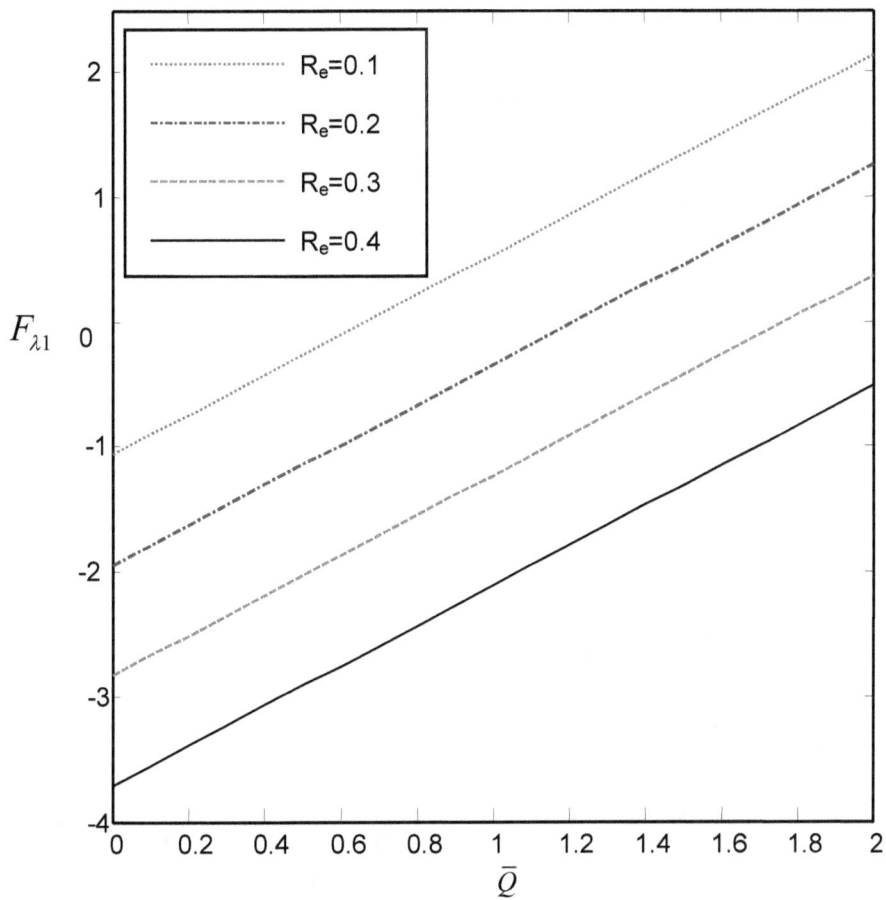

Figure 5.21: Various of $\overline{Q}$ with $F_{\lambda 1}$ at $y=h_1$ for different values of Reynolds R_e with a=0.2, b=1.2, d=1, M=1, $\phi = \pi / 6$, $Da = 0.1$, $\alpha = \pi/6$ and $F_r = 0.1$.

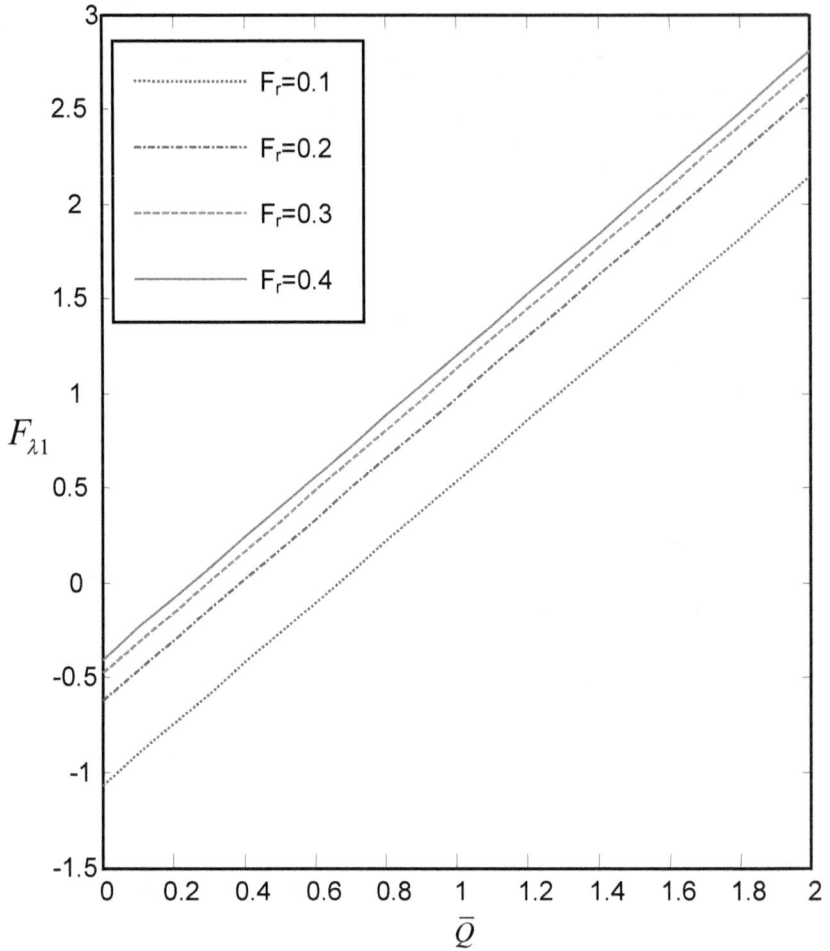

Figure 5.22: Various of $\overline{Q}$ with $F_{\lambda 1}$ at $y=h_1$ for different values of Froude number F_r with a=0.2, b=1.2, d=1, M=1, $\phi = \pi / 6$, Da =0.1, α =π/6 and R_e =0.1.

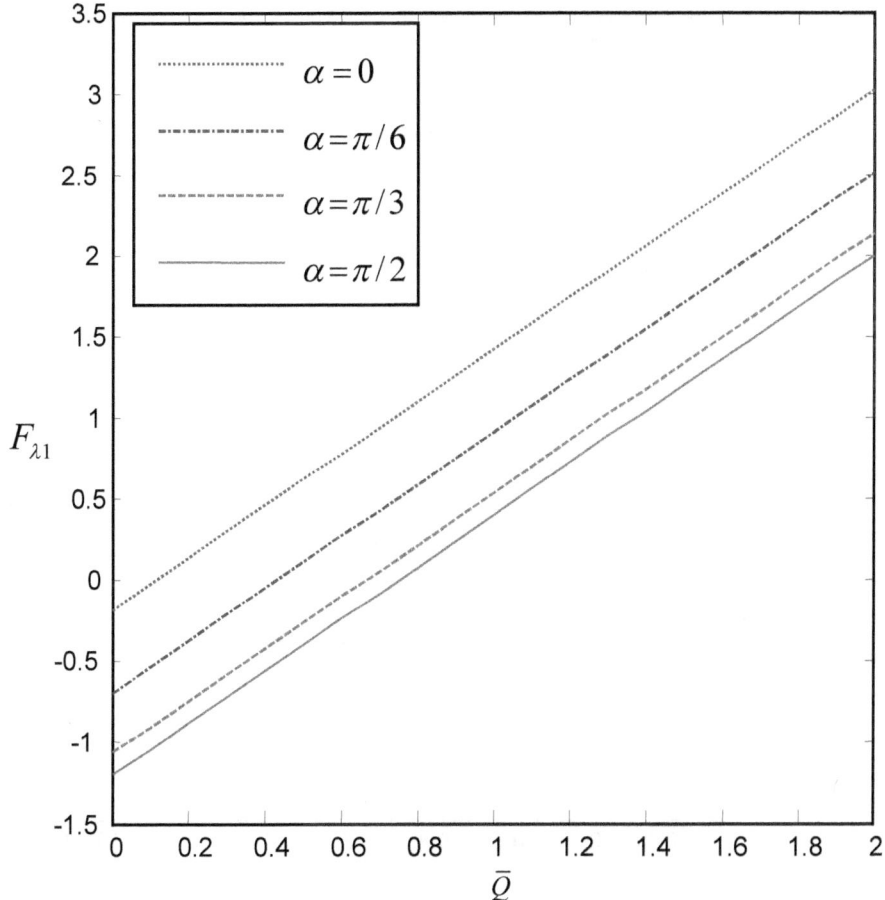

Figure 5.23: Various of $\overline{Q}$ with $F_{\lambda 1}$ at $y=h_1$ for different values of angle of inclination α with a=0.2, b=1.2, d=1, M=1, $\phi=\pi/6$, Da =0.1, R_e =0.1 and F_r =0.1.

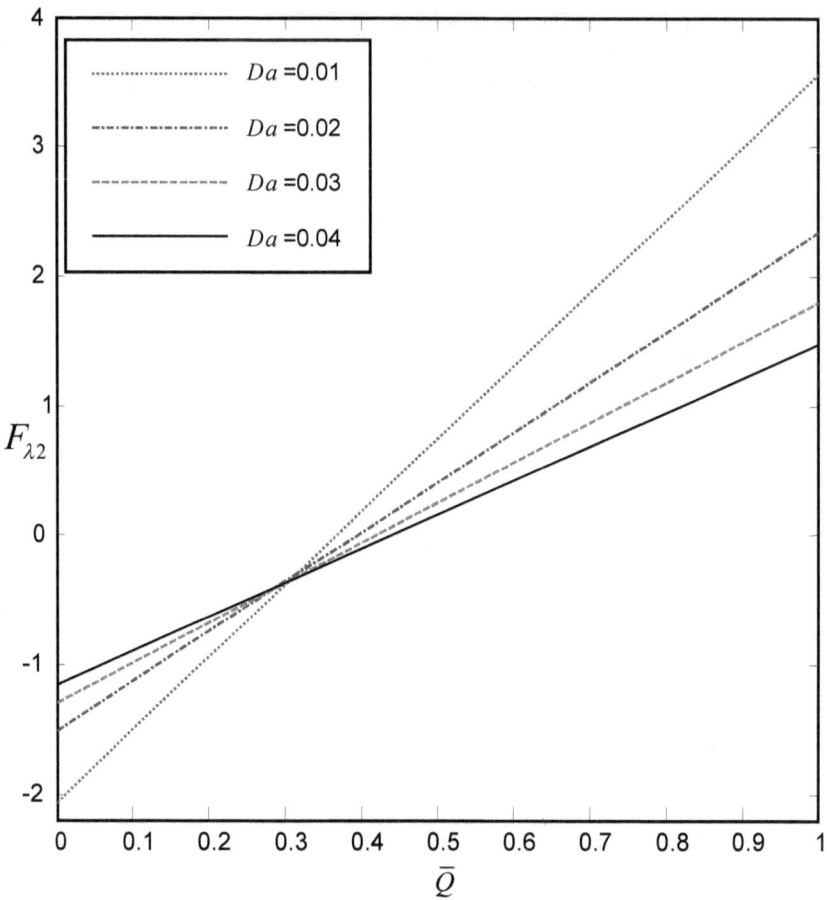

Figure 5.24: Various of $\bar{Q}$ with $F_{\lambda 2}$ at $y = h_2$ for different values of Darcy number Da with a=1.2, b=0.2, d=1, M=1, $\phi = \pi / 6$, R_e =0.1, α =π/4 and F_r =0.1.

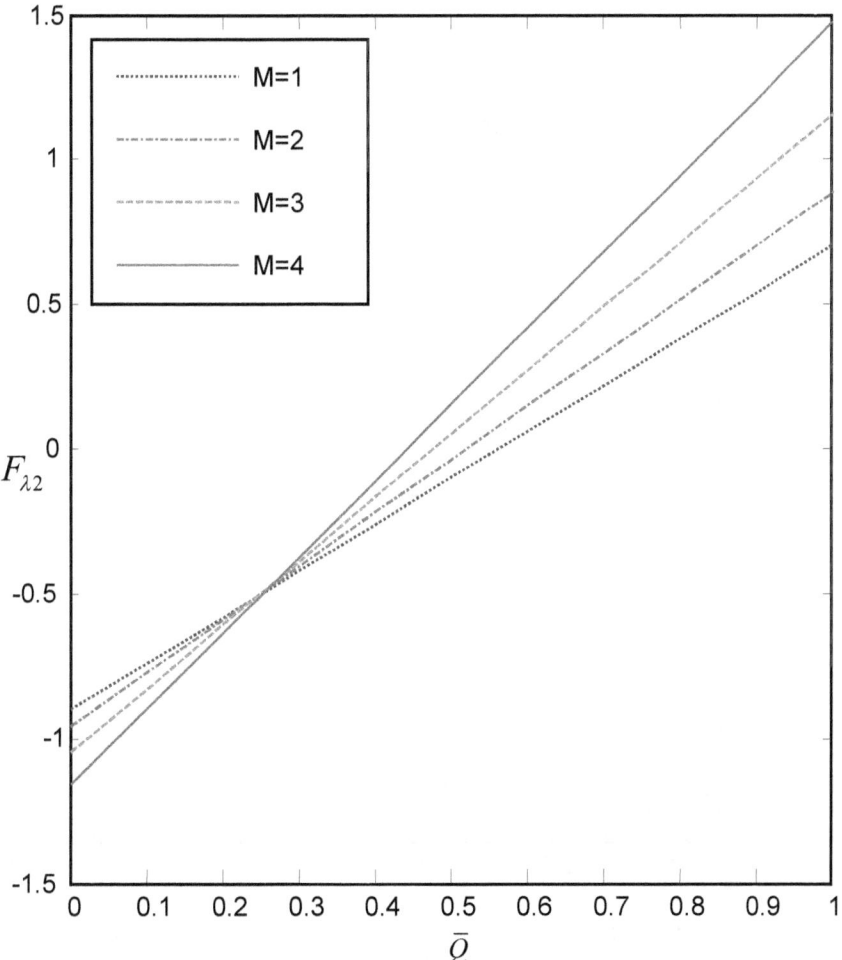

Figure 5.25: Various of $\overline{Q}$ with $F_{\lambda 2}$ at $y = h_2$ for different values of Hartmann

number M with a=1.2, b=0.2, d=1, $\phi = \pi / 6$, Da =0.1, R_e =0.1, α =π/4

and F_r =0.1.

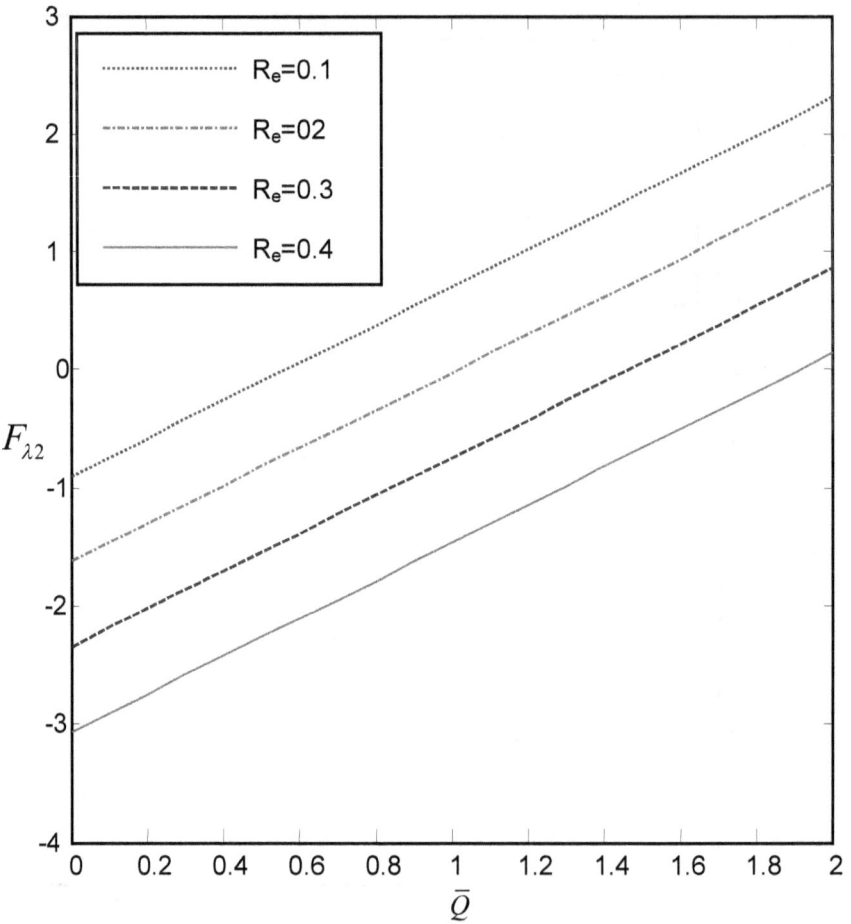

Figure 5.26: Various of $\overline{Q}$ with $F_{\lambda 2}$ at $y = h_2$ for different values of Reynolds

number R_e with a=1.2, b=0.2, d=1, M=1, $\phi = \pi/6$, Da =0.1, α =π/4 and

F_r =0.1.

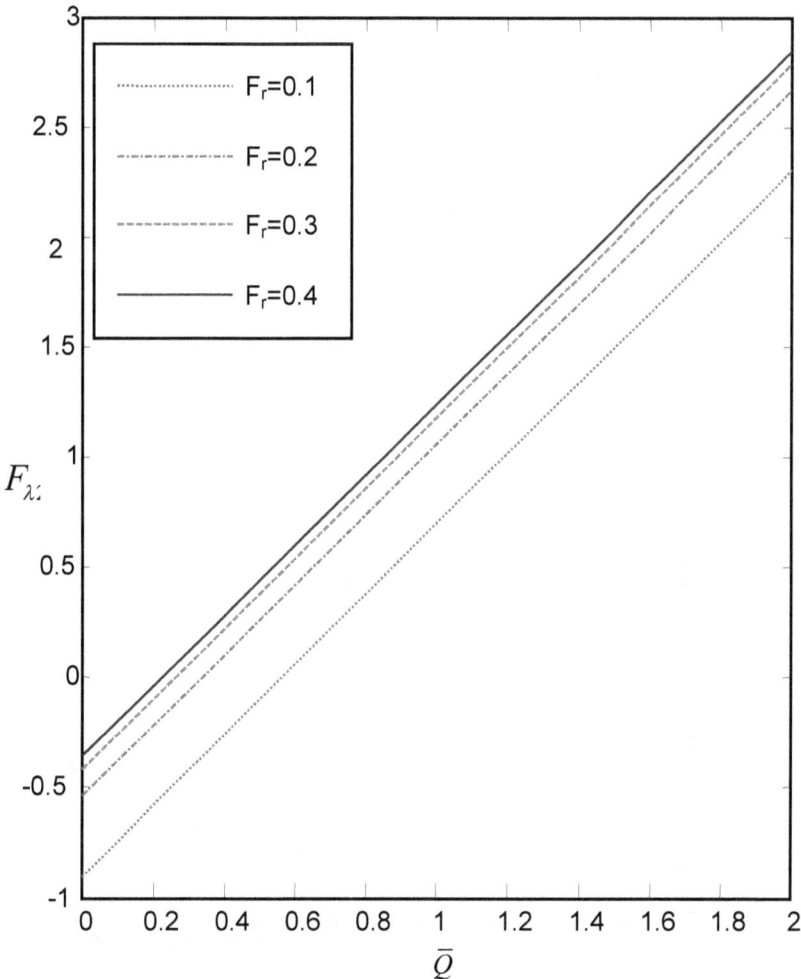

Figure 5.27: Various of $\overline{Q}$ with $F_{\lambda 2}$ at $y = h_2$ for different values of Froude number F_r with a=1.2, b=0.2, d=1, M=1, $\phi = \pi / 6$, Da =0.1, α =π/4 and R_e =0.1.

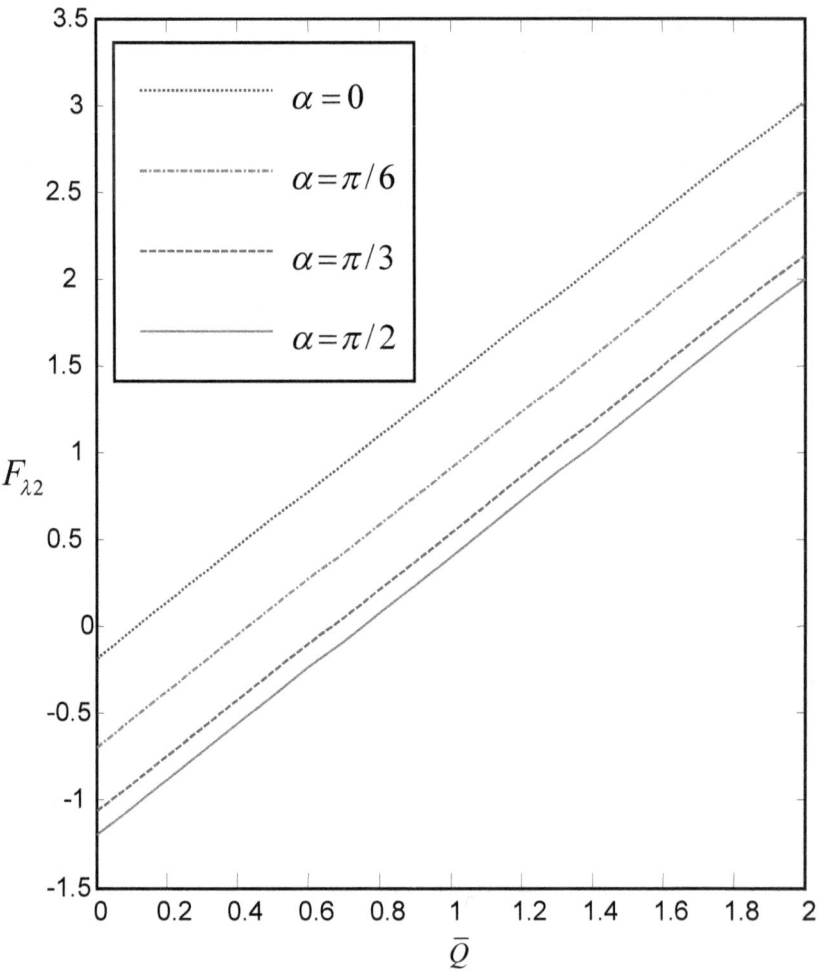

Figure 5.28: Various of $\overline{Q}$ with $F_{\lambda 2}$ at $y = h_2$ for different values of angle of inclination α with a=1.2, b=0.2, d=1, M=1, $\phi = \pi/6$, $Da = 0.1$, $F_r = 0.1$ and $R_e = 0.1$.

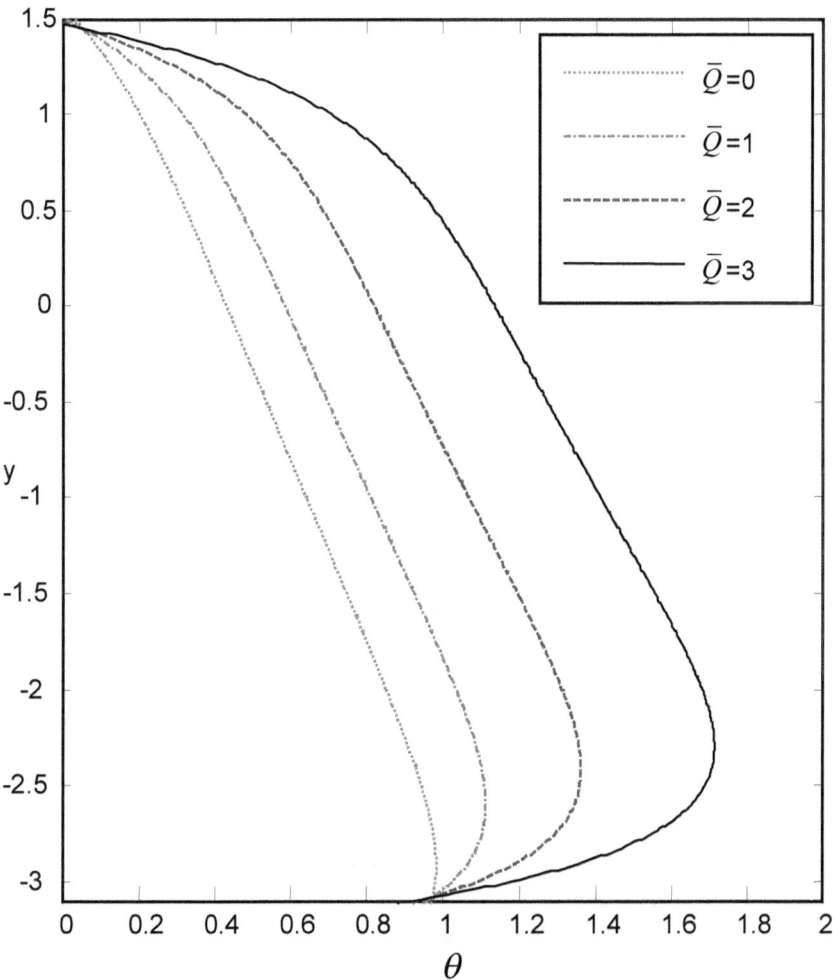

Figure 5.29: Effect of the average flow rate $\overline{Q}$ on temperature distribution θ for
a=0.7, b=1.2, d=2, x=0, ϕ= π/6, M=1, Da =0.1 and Br=1.

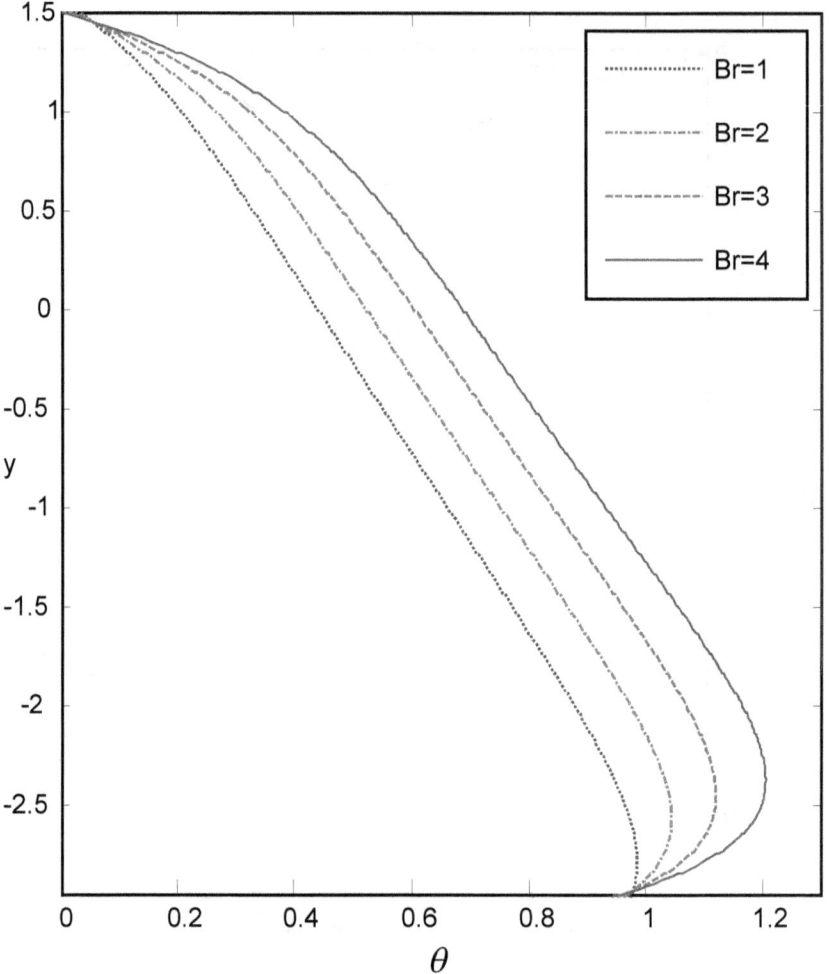

Figure 5.30: Effect of Brinkman number Br on temperature distribution θ for a=0.7,

b=1.2, d=2, x=0, $\overline{Q}$=1, ϕ=π/6, M=1 and Da =0.1

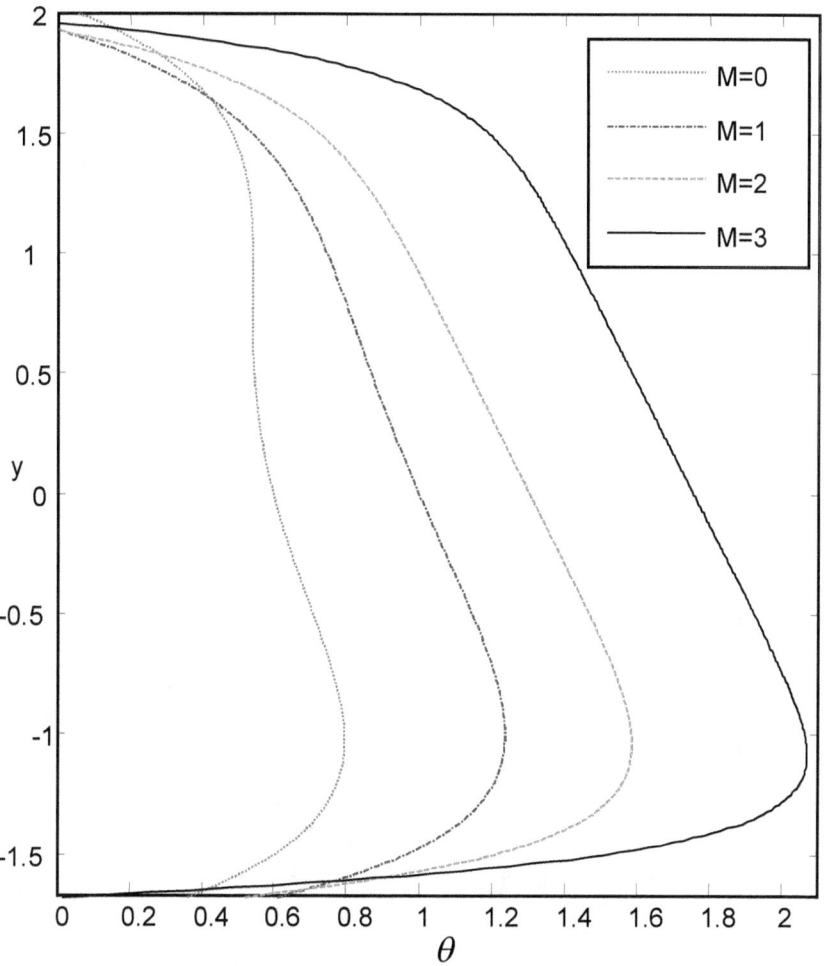

Figure 5.31: Effect of Hartmann number M on temperature distribution θ for a=0.7, b=1.2, d=1, x=0, $\overline{Q}$=1, ϕ=π/3, Da =0.1 and Br=1.

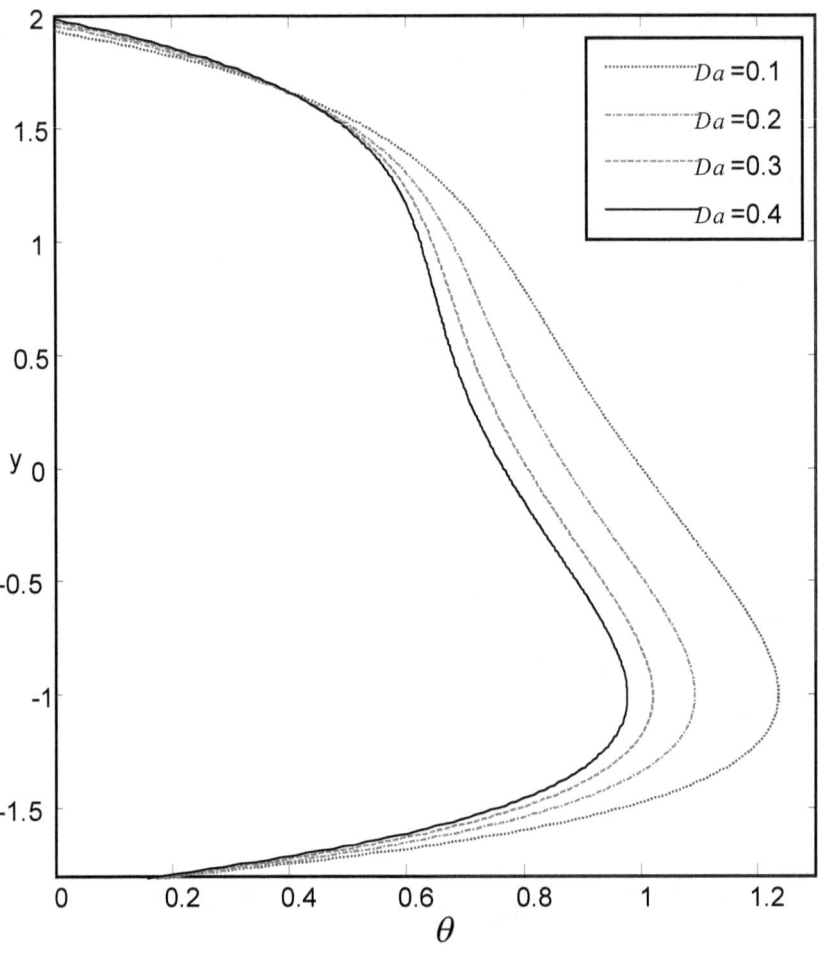

Figure 5.32: Effect of Darcy number Da on temperature distribution θ for a=0.7, b=1.2, d=1, x=0, $\overline{Q}$=1, $\phi=\pi/3$, M=1 and Br=1.

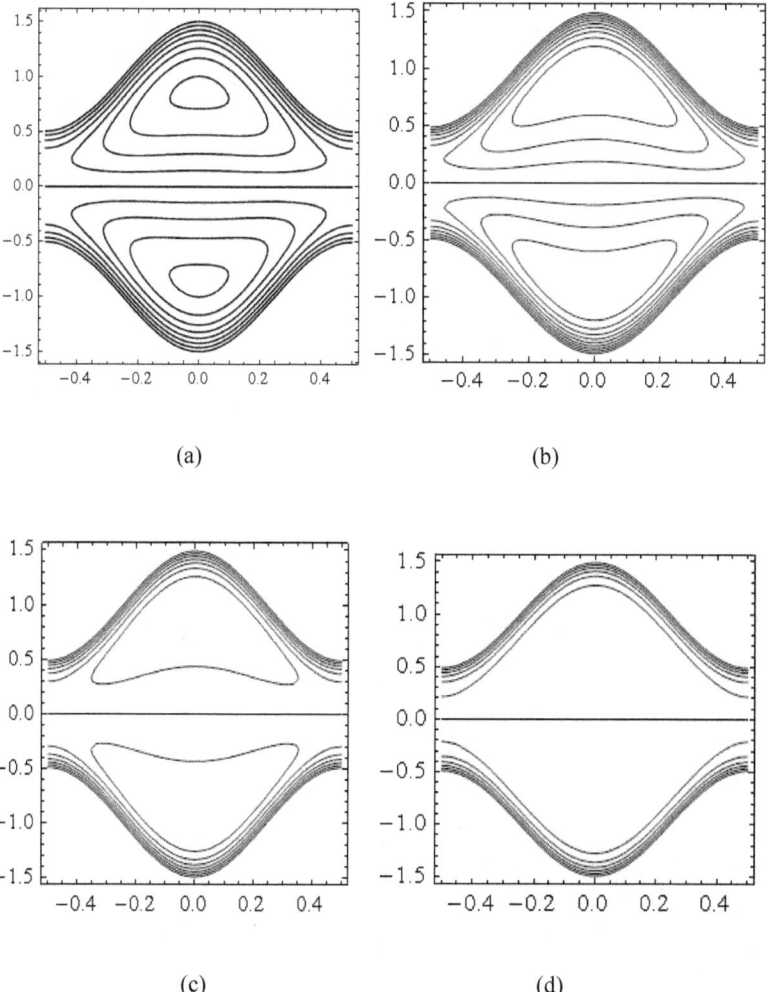

(a) (b)

(c) (d)

Figure 5.33: Stream lines for different values of Hartman number M and fixed values of other parameters chosen as a=0.5, b=0.5, d=1, ϕ=0, $\overline{Q}$=1 and Da =0.2, (a) M=0, (b) M=4, (c) M=6 and (d) M=8.

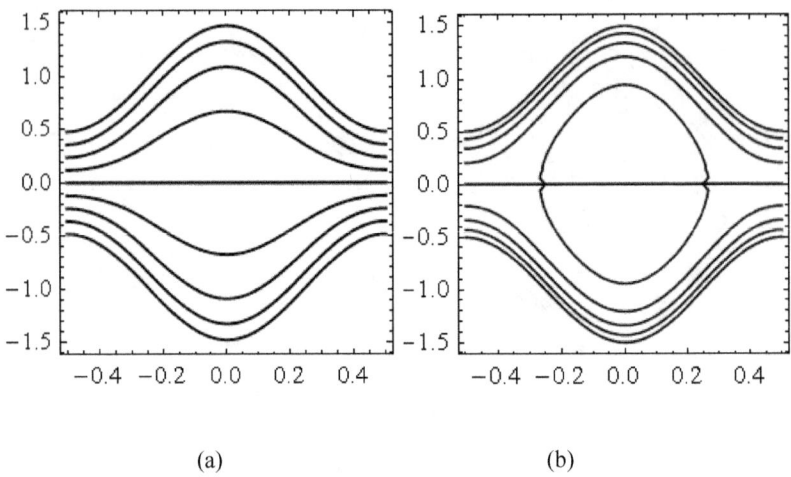

(a)   (b)

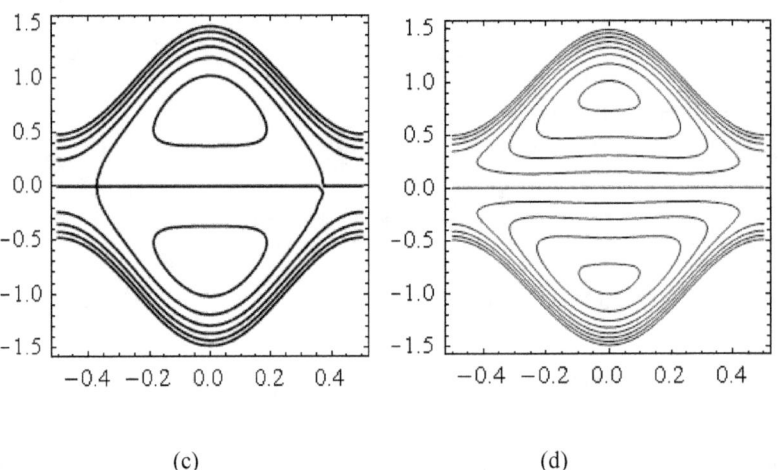

(c)   (d)

Figure 5.34: Stream lines for different values of average volume flow rate $\overline{Q}$ and fixed values of other parameters chosen as a=0.5, b=0.5, d=1, ϕ=0, M=1 and Da=0.1. (a) $\overline{Q}$=1.2, (b) $\overline{Q}$=1.4, (c) $\overline{Q}$=1.6 and (d) $\overline{Q}$=1.8.

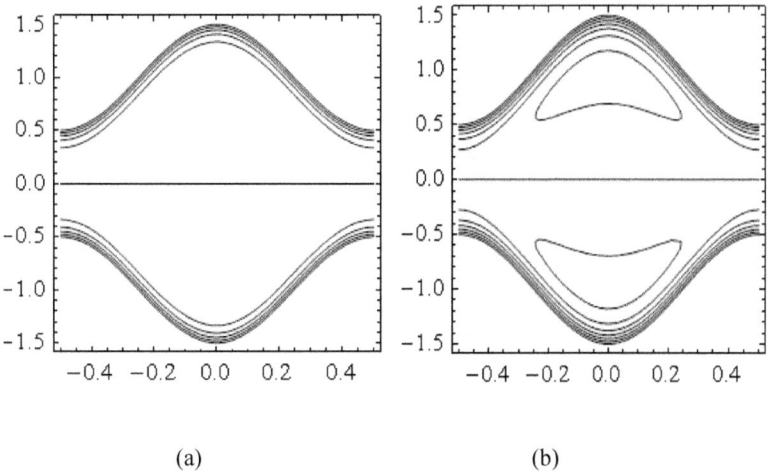

(a)   (b)

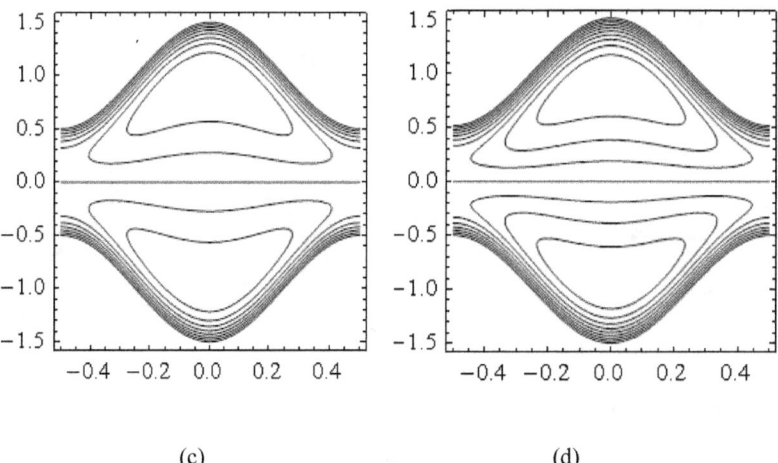

(c)   (d)

Figure 5.35: Stream lines for different values of Darcy number Da and fixed values of other parameters chosen as a=0.5, b=0.5, d=1, ϕ=0, M=1 and $\bar{Q}$=1. (a) Da=0.01, (b) Da=0.02, (c) Da==0.04 and (d) Da=0.06.

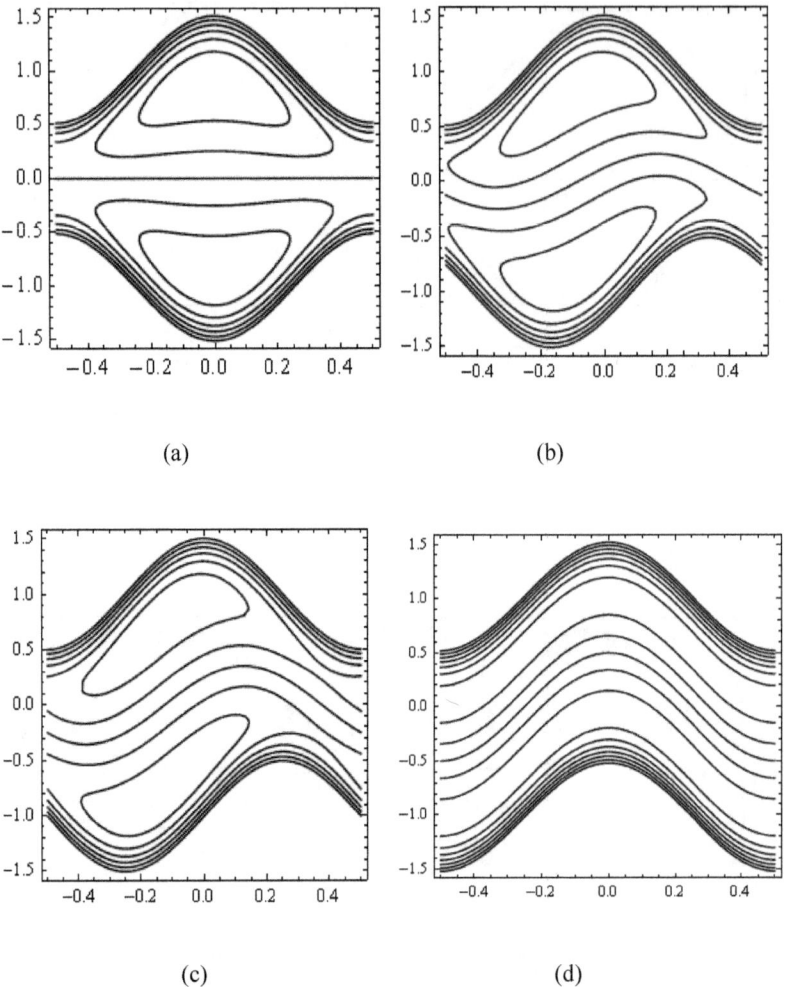

(a)

(b)

(c)

(d)

Figure 5.36: Stream lines for different values of phase difference ϕ and fixed values of other parameters chosen as a=0.5, b=0.5, d=1, Da =0.1, M=1 and $\bar{Q}$ =1, (a) ϕ=0, (b) ϕ= π/3, (c) ϕ= π/2 and (d) ϕ= π.

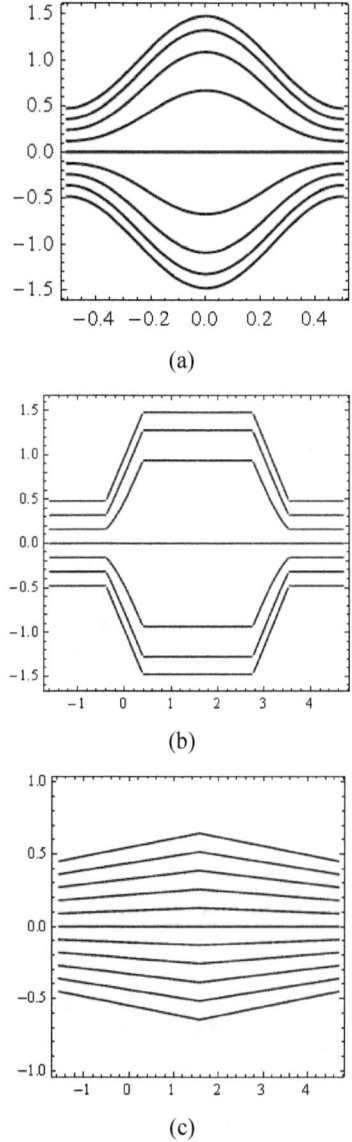

Figure 5.37: Stream lines for different wave forms a) sinusoidal wave b) square wave c) triangular wave with a=0.5, b=0.5, d=1, ϕ=0, Da =0.1, M=1 and $\overline{Q}$ =1,

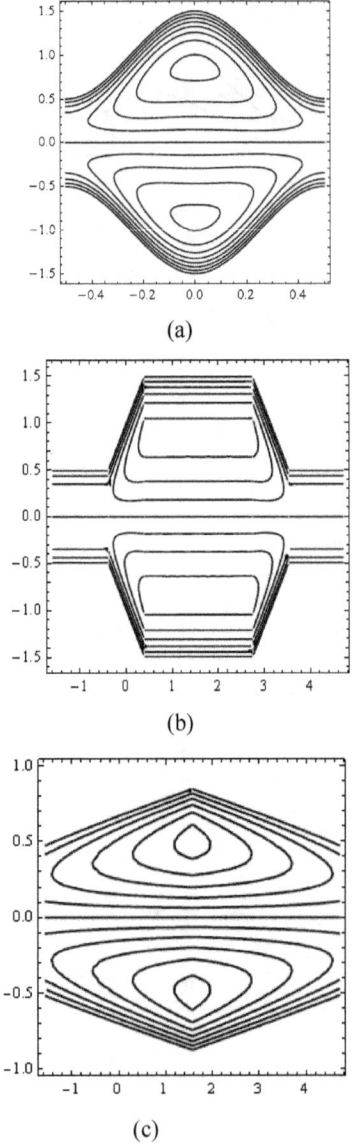

(a)

(b)

(c)

Figure 5.38: Stream lines for different wave forms a) sinusoidal wave, b) square wave and c) triangular wave with a=0.5, b=0.5, d=1, ϕ=0, Da=0.1, M=1 and $\bar{Q}$=1.8.

REFERENCES

ABD ELNAY, M.A. HAROUN, M.H. A new modal for study the development of wall properties on peristaltic transport of a viscous fluid. *Commun Nonlinear Sci Nummer simul* 18(2008), 752-762.

ALI NASIR, HAYAT TASAWAR. Peristaltic motion of a Carreau fluid in an asymmetric channel. *Applied Mathematics and computation* **193** (2007) 535-552.

BEAVERS, G.S. JOSEPH, D.D. Boundary conditions at a naturally permeable wall. *J. Fluid Mech.* **30**(1967), 197-207.

BEST, C.H.AND TAYLOR, N.B. The living body London: Chapman and Hall Ltd, (1958).

BIRD, R.B, STEWART, W.E, LIGHTFOOT, E. N. Transport phenomena, Newyork, wiley (1960).

BRASSEURE, J. G, CORRSIN, S. AND LU, N. Q. The influence of a peripheral layer of different viscosity on peristaltic pumping with Newtonian fluid, *J. Fluid Mech.* **174**(1987), 495-519.

BROWN, T.D. HUNG, T.K. Computational and experimental investigations of two dimensional non linear peristaltic flows. *J. Fluid Mech*, **83**(1977), 249-272.

BRINKMAN, H.E. A Calculation of the viscous force exerted by a flowing fluid in a dense swarm of particles. *Appl. Sci. Res. Al* (1947), 27-37.

BUGLIARELLO, G, SEVILLA, J. Velocity distribution and other characteristic of steady and pulsatile blood flow in time glass tubes. *Biorheology.* **7**(1970), 85-107.

BOYARSKY, S. Surgical physiology of the renal peivis, *Monogr. Surg.* 1 (1964), 173-213.

CHANNABASSAPPA, M.N, RANGANNA, G. *Proc. Indian Acad. Sci.* **83A,** (1976), 145.

DARCY, H.P.G. Les Fontains publicques de la ville de Dijion, paris: victor Dalmost. (1856).

DE VARIES, K. LYONS, EA. BAVARD, J. LEVIS, CS. LINDSAY, DJ. Contraction of the inner third of the myometrium. *Ame. J. Obseterics Gynecol.* **162**(1990), 679-682.

ELSHEHAWAY, E.F. ELDABE, N.T. ELGHAZY, E.M. EBAID, A. Peristaltic transport in an asymmetric channel through a porous medium, *Appl. Math. Comput.* **182**(2006), 140-150.

EYTAN, O. JAFFA, AJ. HAR-TOOV, J. DALACH, E. ELAD, D. Dynamic of the intrauterine fluid-wall interface. *Ann Biomed Eang.* **27**(1999), 372-379.

EYTAN, O. ELAD, D. Analysis of Intra-uterine fluid motion induced uterine contractions. *Bull math Biol.* **61**(1999), 221-238.

EYTAN, O. JAFFA, AJ. HAR-TOOV, J. DALACH, E. ELAD, D. Dynamic of the intrauterine fluid-wall interface. *Ann. Biomed. Eang.* 27(1999), 372-379.

FUNG, Y.C. YIN, C.C. Peristaltic waves in cylindrical tubes. *J. App. Mech.* **36(1969)**, 579-587.

FUNG, Y.C. AND YIH, C.S. Peristaltic transport. *J. Appl. Mech.,* **33**, 669-675.

FUNG, Y. C. AND TANG, H.T. Longitudinal Dispersion of tracer particles in the blood flowing in a pulmonary alveolar sheet, *ASMEJ Appl. Mech.* **42**(1975), 536.

HAYAT, T. NIAZ AHMAD, ALI, N. Effect of an endoscope and magnetic field on the peristaltic involving Jeffrey fluid. *Nonlinear Science and Numerical Solution,* **13**(2008), 1581-1591.

HAYAT, T. KHAN, M. ASGHAR, S. SIDDIQUI, A.M. A mathematical modal of peristalsis in tubes through a porous medium, *J. Porous Media,* **9**(2006), 55-67.

HAYAT, T. MAHOMED, F.M. ASGHAR, S. Peristaltic flow of magnetohydrodynamic Johnson-Segalman fluid, *Nonlinear Dynam.* **40**(2005), 375-385.

HAYAT, T. ALI, N. ASHGHAR, S. Hall effects on peristaltic flow of a Maxwell fluid in a porous medium. *Phys. Celt.* (2007), 363-397.

HAYAT, T. NASIR ALI, ZAHEER ABBAS. Peristaltic flow of a mocropolar fluid in a channel with different wave forms, *Phys. Lett. A* **370**(2008), 331-344.

HAYAT, T. QURESHI, M.V. HUSAIN, Q. Effect of heat transfer on the peristaltic flow of an electrically conducting fluid in a porous space. *Appl. Math. Modelling.***33**(2009), 1862-1873.

HOLMES M.H. Lai W.M. and MOW V.C.. Singular perturbation analysis of the nonlinear, flow depondent compressive study relaxation behavior of articular cartilage, *ASME. J. Biomech Eng* **107**(1985), 206-218.

HUNG, T.K. BROWN, T.D. Solid-particle motion in two dimensional peristaltic flows. *J. Fluid Mech.* **73**(1976), 77-96.

IRENE , GIAMBATTISTA. Perturbation solution for pulsatile flow of a non-Newtonian Williamson fluid in rock fracture. *Mech Mining Sci.* **44**(2007), 271-278.

JAFFRIN, M.Y. AND SHAPIRO, A.H. Peristaltic pumping. *Ann. Rev. Fluid Mech.* **3**(1971), 13-36.

JAYARAMAN, G. AND JAIN, R. A theoretical models for water flux through the artery–wall. *J. Biomech, Engng* **109** (1987), 311- 317.

JAYARAMAN, G. Water transport in the arterial wall – A theoretical study. *J. Biomechanics* **16**(1983), 833- 840.

KAPUR, J.N. Mathematical Modeling in Biology and Medicine, *Affiliated East West Press, Pvt. Ltd* (1985).

KENYON, D.E. The theory of an incompressible solid fluid mixture. *Arch. Rat. Mech Anal.* **62**(1976), 131- 147.

KENYON, D.E. A mathematical model of water flux through arotic tissue. *Bull, Math, Biology,* **41**(1979), 70- 79.

KOTHANDAPANI, M. SRINIVAS, S. Peristaltic transport of a Jeffrey fluid under the effect of magnetic field in an asymmetric channel. *Int J Non-linear Mech.***43** (2008), 915-924.

KOTHANDAPANI, M. SRINIVAS, S. On the influence of wall properties in the MHD peristaltic transport with heat transfer and porous medium. *Psys. Lett.* **A372(2008)** 4586-4591.

KILL, F. The function of the ureter and the renal pelvies. Saunders, philadephia (1957).

KWAN, M.K. Finite deformational theory for a non linearly permeably biphasic medium, *Ph.D thesis, Renssecular polytechnic institute* (1985).

LATHAM, T.W. Fluid motions in a peristaltic pump. *M.S. Thesis, M.I.T.* (1966).

LU, N.Q. The influence of two Newtonian fluid with different viscosity on the peristaltic pumping. *An essay submitted to the Johna Hopkins University for the degree M. S. Baltimore, Maryland* (1975).

MEKHEIMER, KH. S. Non-linear peristaltic transport through a porous medium in an inclined planner channel. *J. Porous Media* 6(2003), 189.

MEKHEIMER, KH. S. AL-ARABI, T.H. Non-Linear peristaltic of MHD flow through a porous medium. *Int. J. Math. Sci.* 26(2003), 1663.

MEKHEIMER, KH. S. Effect of magnetic field on peristaltic flow of a couple stress fluid. *Phys. Lett.* A 371(2008), 4271-4278.

MISHRA, J.C. AND GHOSH, S.K. A Mathematical model for the study of blood flow through a channel with permeable walls, *Acta Mech.* 122(1997), 137-153.

MISHRA, M. RAO, A.R. Nonlinear and curvature effects on peristaltic flow of a viscous fluid in an asymmetric channel, *Acta Mechanica.* 168(2004), 35-39.

MISHRA, M. RAO, A. R. Peristaltic transport of a Newtonian fluid in an asymmetric channel, *Z. Angew. Math. Phys.* 54(2004), 532-550.

MOW, V.C. HOLMES, M. H. and LAI, W.M. Fluid transport and mechanical properties of articular cartilage. *A review. J. Biomechanical.* 17(1984), 377-397.

NADEEM, S. AKRAM, S. Peristaltic flow of a Williamson fluid in an asymmetric channel. *Commun Nonlinear Sci Numer Simulate* 15(2010) 1705-1716.

NEREM, R.M AND LEVESQUE, M.J. Hand Book of Bio Engineering. *Mc. Graw Hill*, Newyork (1987).

PRASANNA HARIHARAN. SESHADRI, V. RUPAK K. BANERJEE. Peristaltic transport of non-Newtonian fluid in a diverging tube with different wave forms. *Mathematical and Computer Modeling.* 48(2008), 998-1017.

POZRIKIDIS, C. A study of peristaltic flow. *J Fluid Mech*, 180(1987), 515-527.

RAJU, K.K. DEVANATHAN. Peristaltic motion of a non-Newtonian fluid. *Acta* 11(1972), 170-178.

RAO, A.R. AND USHA, S. Peristaltic transport of two immiscible viscous fluids in a circular cylinder tube. *J. Fluid Mech.* 298(1995), 271-285.

RUDRAIAH, G.W. and VEERABHADRAIAH, *R.J. Math and phys. Sci.*13(6) (1979).523.

RUDRAIAH, N and WILFRED. V. *Nat. Acad. Sci. Lett,* 49(1980).

SAFFMAN, P.G. On the boundary condition at the surface of porous medium. *Studies in Appl. Maths,* L. 93(1971).

SCOTT BLAIR, G.W. An equation for the flow of blood, plasma and serum through glass capillaries, *Nature* 183(1959), 631.

SCOTTBLAIR, G.W. AND SPNNER, D.C. An introduction to Biorheology. *Elsevier scientific publishing company, Amsterdam*(1974).

SEMANS, J.H. LONGWORTH, O.R. Morphology motility and fertility of spermatozoa recovered from different areas of ligated rabbit epiddymis. J. *Report Fert. IT.* (1935), 125-137.

SHAPIRO, A.H. JAFFRIN, M.Y. AND WEINBERG, SL. Peristaltic pumping with long wavelength at low Reynolds number. *J. Fluid Mech.* 37(1969) 799-825.

SHIVAKUMAR, P.N., NAGARAJ, S., VEERABHADRAIAH, R. & RUDRAIAH, N. Fluid movement in a channel of varying gap with permeable walls covered by porous media. *Int. Engg. Sci.* 24:4(1986), 479-492.

SHUKLA, J.B. GUPTA, S.P. Peristaltic transport of a powerlaw fluid with variable consistency, *J. Biomech. Eng.* 104. (1982), 182-186.

SIDDIQUI, A. M. HAYAT, T. KHAN MASOOD. Magnetic fluid model induced by peristaltic waves, *Journal of Physical Society of Japan.* 73(2004), 2142-2147.

SRIVASTAVA, L.M. ASGHAR, A.M. Oscillating flow of a conducting fluid with suspension of spherical particles, J. Appl. Mech. 47(1980) 196-199.

SRIVASTAVA, L.M. AND SRIVASTAVA, V.P. Peristaltic transport of blood: Casson model II, *J. Fluid. Mech.* 122(1984), 439-465.

SRINIVAS, S. KOTHANDAPANI, M. Peristaltic transport in an asymmetric channel with heat transfer. *Int. Commun. Heat Transfer* **35**(2008), 514-522.

STUD, V.K. SEPHONE, G.S. MISHRA, R.K.G. Pumping action blood flow by a magnetic field. *Bull. Biol.* 39(1977), 358-390.

SUBBA REDDY, M.V. SREENADH, S. RAMCHANDRA RAO, A. Peristaltic motion of a power-law fluid in an asymmetric channel. *Int. J. Non-Linear Mech.* **42**(2007), 1153-1161.

TAKABATAKE, S. AYUKKAWA, K. Numerical study of two-dimensional peristaltic flow. *J. Fluid. Mech.* **122**(1982) 439-465.

TANG, D. AND RANKIN, S. Numerical and asymptotic solution for peristaltic motion of nonlinear viscous flows with elastic free boundaries. *SIAM. J. Sci. Comput.* **14**(1993), 1300-1319.

TANG, H.T. AND FUNG, Y.C. Fluid movement in a channel with permeable walls covered by porous media: A model of lung alveolar sheet. *J. Appl. Mech.* **97.** (1975), 45-50.

USHA, S. & RAO, A.R. Peristaltic transport of two-layered power-law fluids. *J. Biomech. Engg.* **119.** (1997), 483-488.

VAJRAVELU, K. SREENADH, S. RAMAKRISHNA, S. AND ARUNACHALM, P.V. Bingham fluid flow through a circular pipe with permeable wall. *ZAMM.* **07**(1987), 568-569.

VAJRAVELU, K., SREENADH, S. AND ARUNACHALM, P.V. Unsteady flow of two immiscible conducting fluids between two permeable beds. *Journal of mathematical Analysis and Applications.* **196**(1995), 1105-1116.

VAJRAVELU, K., SREENADH, S. and ARUNCHALAM, P.V. Combined free and forced convection in an inclined channel with permeable boundaries. *Journal of math. Anal.and Appl vol.* **166** (1992), 2393-403.

VAJRAVELU, K. SREENADH, S. AND RAMESH BABU, V. Peristaltic pumping of a Herschel-Bulkley fluid in a channel. *Appl. Math. And compute,* **169**(2005a), 729-735.

VAJRAVELU, K. SREENADH, S. AND RAMESH BABU, V. Peristaltic pumping of a Herschel-Bulkley fluid in an inclined tube, *International Journal of Nonlinear Mechanics*, **40**(2005b), **83**.

VAJRAVELU, K. SREENADH, S. AND RAMESH BABU, V. Peristaltic transport of a Herschel-Bulkely fluid in contact with a Newtonian fluid, *Quarterly. J. of Appl. Math, LXIV,* **4**(2006), 593-604.

VAJRAVELU, KUPPALAPALLE. SREENADH, S. HEMADRI REDDY, R. MURUGESAN, K. Peristaltic transport of a Casson fluid in contact with a Newtonian Fluid in a circular Tube with permeable wall, *International Journal of Fluid Mechanics Research, Vol.* **36**(2009)(3), 244-254.

WANG, Y. HAYAT, T. ALI, N. OBERLACK, M. Magnetohydrodynamics peristaltic motion of Sisko fluid in a symmetric or asymmetric channel. *Physica* **A 387**(2008), 347-362.

WAZWAZ, A.M. Adomian decomposition method for a reliable treatment of the Emden-fowler equation. *Appl. Math. Comput.* **161**(2005), 543-560.

WEINBERG, S.L. ECKISTEIN, E.C. AND SHAPIRO, A.H. An experimental study of study of peristaltic pumping, *J. Fluid Mech.* **49**(1971), 461-497.

YIN, C.C. AND FUNG, Y.C. Comparison of theory and experimental in peristaltic transport, *J. Fluid Mech.* **47**(1971), 93-122.